Mastering Spark with R
The Complete Guide to Large-Scale Analysis and Modeling

Javier Luraschi, Kevin Kuo, and Edgar Ruiz

Beijing · Boston · Farnham · Sebastopol · Tokyo

Mastering Spark with R

by Javier Luraschi, Kevin Kuo, and Edgar Ruiz

Copyright © 2020 Javier Luraschi, Kevin Kuo, and Edgar Ruiz. All rights reserved.

Published by O'Reilly Media, Inc., 1005 Gravenstein Highway North, Sebastopol, CA 95472.

O'Reilly books may be purchased for educational, business, or sales promotional use. Online editions are also available for most titles (*http://oreilly.com*). For more information, contact our corporate/institutional sales department: 800-998-9938 or *corporate@oreilly.com*.

Acquisition Editor: Jonathan Hassell	**Indexer:** Judy McConville
Development Editor: Melissa Potter	**Interior Designer:** David Futato
Production Editor: Elizabeth Kelly	**Cover Designer:** Karen Montgomery
Copyeditor: Octal Publishing, LLC	**Illustrator:** Rebecca Demarest
Proofreader: Rachel Monaghan	

October 2019:   First Edition

Revision History for the First Release

2019-10-04:   First Release

See *http://oreilly.com/catalog/errata.csp?isbn=9781492046370* for release details.

978-1-492-04637-0

[LSI]

To Adrian, Clara, Julian, Max, Mila and Roman.

Table of Contents

Foreword

Apache Spark is a distributed computing platform built on extensibility: Spark's APIs make it easy to combine input from many data sources and process it using diverse programming languages and algorithms to build a data application. R is one of the most powerful languages for data science and statistics, so it makes a lot of sense to connect R to Spark. Fortunately, R's rich language features enable simple APIs for calling Spark from R that look similar to running R on local data sources. With a bit of background about both systems, you will be able to invoke massive computations in Spark or run your R code in parallel from the comfort of your favorite R programming environment.

This book explores using Spark from R in detail, focusing on the `sparklyr` package that enables support for `dplyr` and other packages known to the R community. It covers all of the main use cases in detail, ranging from querying data using the Spark engine to exploratory data analysis, machine learning, parallel execution of R code, and streaming. It also has a self-contained introduction to running Spark and monitoring job execution. The authors are exactly the right people to write about this topic—Javier, Kevin, and Edgar have been involved in `sparklyr` development since the project started. I was excited to see how well they've assembled this clear and focused guide about using Spark with R.

I hope that you enjoy this book and use it to scale up your R workloads and connect them to the capabilities of the broader Spark ecosystem. And because all of the infrastructure here is open source, don't hesitate to give the developers feedback about making these tools better.

— Matei Zaharia
Assistant Professor at Stanford University,
Chief Technologist at Databricks,
and original creator of Apache Spark

Preface

In a world where information is growing exponentially, leading tools like Apache Spark provide support to solve many of the relevant problems we face today. From companies looking for ways to improve based on data-driven decisions, to research organizations solving problems in health care, finance, education, and energy, Spark enables analyzing much more information faster and more reliably than ever before.

Various books have been written for learning Apache Spark; for instance, *Spark: The Definitive Guide* (*https://oreil.ly/gMaGP*) is a comprehensive resource, and *Learning Spark* (*https://oreil.ly/1-4CA*) is an introductory book meant to help users get up and running (both are from O'Reilly). However, as of this writing, there is neither a book to learn Apache Spark using the R computing language nor a book specifically designed for the R user or the aspiring R user.

There are some resources online to learn Apache Spark with R, most notably the spark.rstudio.com (*https://spark.rstudio.com*) site and the Spark documentation site at spark.apache.org (*http://bit.ly/31H2nMl*). Both sites are great online resources; however, the content is not intended to be read from start to finish and assumes you, the reader, have some knowledge of Apache Spark, R, and cluster computing.

The goal of this book is to help anyone get started with Apache Spark using R. Additionally, because the R programming language was created to simplify data analysis, it is also our belief that this book provides the easiest path for you to learn the tools used to solve data analysis problems with Spark. The first chapters provide an introduction to help anyone get up to speed with these concepts and present the tools required to work on these problems on your own computer. We then quickly ramp up to relevant data science topics, cluster computing, and advanced topics that should interest even the most experienced users.

Therefore, this book is intended to be a useful resource for a wide range of users, from beginners curious to learn Apache Spark, to experienced readers seeking to understand why and how to use Apache Spark from R.

This book has the following general outline:

Introduction
> In the first two chapters, Chapter 1, *Introduction*, and Chapter 2, *Getting Started*, you learn about Apache Spark, R and the tools to perform data analysis with Spark and R.

Analysis
> In Chapter 3, *Analysis*, you learn how to analyze, explore, transform, and visualize data in Apache Spark with R.

Modeling
> In the Chapter 4, *Modeling* and Chapter 5, *Pipelines*, you learn how to create statistical models with the purpose of extracting information, predicticting outcomes, and automating this process in production-ready workflows.

Scaling
> Up to this point, the book has focused on performing operations on your personal computer and with limited data formats. Chapter 6, *Clusters*, Chapter 7, *Connections*, Chapter 8, *Data*, and Chapter 9, *Tuning*, introduce distributed computing techniques required to perform analysis and modeling across many machines and data formats to tackle the large-scale data and computation problems for which Apache Spark was designed.

Extensions
> Chapter 10, *Extensions*, describes optional components and extended functionality applicable to specific, relevant use cases. You learn about alternative modeling frameworks, graph processing, preprocessing data for deep learning, geospatial analysis, and genomics at scale.

Advanced
> The book closes with a set of advanced chapters, Chapter 11, *Distributed R*, Chapter 12, *Streaming*, and Chapter 13, *Contributing*; these will be of greatest interest to advanced users. However, by the time you reach this section, the content won't seem as intimidating; instead, these chapters will be equally relevant, useful, and interesting as the previous ones.

The first group of chapters, 1–5, provides a gentle introduction to performing data science and machine learning at scale. If you are planning to read this book while also following along with code examples, these are great chapters to consider executing the code line by line. Because these chapters teach all of the concepts using your personal computer, you won't be taking advantage of multiple computers, which Spark was designed to use. But worry not: the next set of chapters will teach this in detail!

The second group of chapters, 6–9, introduces fundamental concepts in the exciting world of cluster computing using Spark. To be honest, they also introduce some of the not-so-fun parts of cluster computing, but believe us, it's worth learning the concepts we present. Besides, the overview sections in each chapter are especially interesting, informative, and easy to read, and help you develop intuitions as to how cluster computing truly works. For these chapters, we actually don't recommend executing the code line by line—especially not if you are trying to learn Spark from start to finish. You can always come back and execute code after you have a proper Spark cluster. If you already have a cluster at work or you are really motivated to get one, however, you might want to use Chapter 6 to pick one and then Chapter 7 to connect to it.

The third group of chapters, 10–13, presents tools that should be quite interesting to most readers and will make it easier to follow along. Many advanced topics are presented, and it is natural to be more interested in some topics than others; for instance, you might be interested in analyzing geographic datasets, or perhaps you're more interested in processing real-time datasets, or maybe you'd like to do both! Based on your personal interests or problems at hand, we encourage you to execute the code examples that are most relevant to you. All of the code in these chapters is written to be executed on your personal computer, but you are also encouraged to use proper Spark clusters given that you'll have the tools required to troubleshoot issues and tune large-scale computations.

Formatting

Tables generated from code are formatted as follows:

```
# A tibble: 3 x 2
  numbers text
    <dbl> <chr>
1       1 one
2       2 two
3       3 three
```

The dimensions of the table (number of rows and columns) are described in the first row, followed by column names in the second row and column types in the third row. There are also various subtle visual improvements provided by the `tibble` package that we make use of throughout this book.

Most plots are rendered using the `ggplot2` package and a custom theme available in the appendix; however, because this book is not focused on data visualization, we only provide code to render a basic plot that won't match the formatting we applied. If you are interested in learning more about visualization in R, consider specialized books like *R Graphics Cookbook* (*https://oreil.ly/bIF4l*) (O'Reilly).

Acknowledgments

We thank the package authors that enabled Spark with R: Javier Luraschi, Kevin Kuo, Kevin Ushey, and JJ Allaire (`sparklyr`); Romain François and Hadley Wickham (`dbplyr`); Hadley Wickham and Edgar Ruiz (`dpblyr`); Kirill Mülller (`DBI`); and the authors of the Apache Spark project itself, and its original author Matei Zaharia.

We thank the package authors that released extensions to enrich the Spark and R ecosystem: Akhil Nair (`crassy`); Harry Zhu (`geospark`); Kevin Kuo (`graphframes`, `mleap`, `sparktf`, and `sparkxgb`); Jakub Hava, Navdeep Gill, Erin LeDell, and Michal Malohlava (`rsparkling`); Jan Wijffels (`spark.sas7bdat`); Aki Ariga (`sparkavro`); Martin Studer (`sparkbq`); Matt Pollock (`sparklyr.nested`); Nathan Eastwood (`sparkts`); and Samuel Macêdo (`variantspark`).

We thank our wonderful editor, Melissa Potter, for providing us with guidance, encouragement, and countless hours of detailed feedback to make this book the best we could have ever written.

To Bradley Boehmke, Bryan Adams, Bryan Jonas, Dusty Turner, and Hossein Falaki, we thank you for your technical reviews, time, and candid feedback, and for sharing your expertise with us. Many readers will have a much more pleasant experience thanks to you.

Thanks to RStudio, JJ Allaire, and Tareef Kawaf for supporting this work, and the R community itself for its continuous support and encouragement.

Max Kuhn, thank you for your invaluable feedback on Chapter 4, in which, with his permission, we adapted examples from his wonderful book *Feature Engineering and Selection: A Practical Approach for Predictive Models* (CRC Press).

We also thank everyone indirectly involved but not explicitly listed in this section; we are truly standing on the shoulders of giants.

This book itself was written in R using `bookdown` by Yihui Xie, `rmarkdown` by JJ Allaire and Yihui Xie, and `knitr` by Yihui Xie; we drew the visualizations using `ggplot2` by Hadley Wickham and Winston Chang; we created the diagrams using `nomnoml` by Daniel Kallin and Javier Luraschi; and we did the document conversions using `pandoc` by John MacFarlane.

Conventions Used in This Book

The following typographical conventions are used in this book:

Italic
> Indicates new terms, URLs, email addresses, filenames, and file extensions.

`Constant width`

Used for program listings as well as within paragraphs to refer to program elements such as variable or function names, databases, data types, environment variables, statements, and keywords.

`Constant width bold`

Shows commands or other text that should be typed literally by the user.

`Constant width italic`

Shows text that should be replaced with user-supplied values or by values determined by context.

 This element signifies a tip or suggestion.

 This element signifies a general note.

Using Code Examples

Supplemental material (code examples, exercises, etc.) is available for download at *https://github.com/r-spark/the-r-in-spark*.

This book is here to help you get your job done. In general, if example code is offered with this book, you may use it in your programs and documentation. You do not need to contact us for permission unless you're reproducing a significant portion of the code. For example, writing a program that uses several chunks of code from this book does not require permission. Selling or distributing a CD-ROM of examples from O'Reilly books does require permission. Answering a question by citing this book and quoting example code does not require permission. Incorporating a significant amount of example code from this book into your product's documentation does require permission.

We appreciate, but do not require, attribution. An attribution usually includes the title, author, publisher, and ISBN. For example: "*Mastering Spark with R* by Javier Luraschi, Kevin Kuo, and Edgar Ruiz (O'Reilly). Copyright 2020 Javier Luraschi, Kevin Kuo, and Edgar Ruiz, 978-1-492-04637-0."

If you feel your use of code examples falls outside fair use or the permission given above, feel free to contact us at *permissions@oreilly.com*.

O'Reilly Online Learning

 For almost 40 years, *O'Reilly Media* has provided technology and business training, knowledge, and insight to help companies succeed.

Our unique network of experts and innovators share their knowledge and expertise through books, articles, conferences, and our online learning platform. O'Reilly's online learning platform gives you on-demand access to live training courses, in-depth learning paths, interactive coding environments, and a vast collection of text and video from O'Reilly and 200+ other publishers. For more information, please visit *http://oreilly.com*.

How to Contact Us

Please address comments and questions concerning this book to the publisher:

O'Reilly Media, Inc.
1005 Gravenstein Highway North
Sebastopol, CA 95472
800-998-9938 (in the United States or Canada)
707-829-0515 (international or local)
707-829-0104 (fax)

We have a web page for this book, where we list errata, examples, and any additional information. You can access this page at *https://oreil.ly/SparkwithR*.

To comment or ask technical questions about this book, send email to *bookques-tions@oreilly.com*.

For more information about our books, courses, conferences, and news, see our website at *http://www.oreilly.com*.

Find us on Facebook: *http://facebook.com/oreilly*

Follow us on Twitter: *http://twitter.com/oreillymedia*

Watch us on YouTube: *http://www.youtube.com/oreillymedia*

CHAPTER 1
Introduction

You know nothing, Jon Snow.

—Ygritte

With information growing at exponential rates, it's no surprise that historians are referring to this period of history as the Information Age. The increasing speed at which data is being collected has created new opportunities and is certainly poised to create even more. This chapter presents the tools that have been used to solve large-scale data challenges. First, it introduces Apache Spark as a leading tool that is democratizing our ability to process large datasets. With this as a backdrop, we introduce the R computing language, which was specifically designed to simplify data analysis. Finally, this leads us to introduce `sparklyr`, a project merging R and Spark into a powerful tool that is easily accessible to all.

Chapter 2, *Getting Started* presents the prerequisites, tools, and steps you need to perform to get Spark and R working on your personal computer. You will learn how to install and initialize Spark, get introduced to common operations, and get your very first data processing and modeling task done. It is the goal of that chapter to help anyone grasp the concepts and tools required to start tackling large-scale data challenges which, until recently, were accessible to just a few organizations.

You then move into learning how to analyze large-scale data, followed by building models capable of predicting trends and discover information hidden in vast amounts of information. At which point, you will have the tools required to perform data analysis and modeling at scale. Subsequent chapters help you move away from your local computer into computing clusters required to solve many real world problems. The last chapters present additional topics, like real-time data processing and graph analysis, which you will need to truly master the art of analyzing data at any scale. The last chapter of this book provides you with tools and inspiration to consider contributing back to the Spark and R communities.

We hope that this is a journey you will enjoy, that will help you to solve problems in your professional career, and to nudge the world into making better decisions that can benefit us all.

Overview

As humans, we have been storing, retrieving, manipulating, and communicating information since the Sumerians in Mesopotamia developed writing around 3000 BC. Based on the storage and processing technologies employed, it is possible to distinguish four distinct phases of development: premechanical (3000 BC to 1450 AD), mechanical (1450–1840), electromechanical (1840–1940), and electronic (1940–present).[1]

Mathematician George Stibitz used the word *digital* to describe fast electric pulses back in 1942,[2] and to this day, we describe information stored electronically as digital information. In contrast, *analog* information represents everything we have stored by any nonelectronic means such as handwritten notes, books, newspapers, and so on.

The World Bank report on digital development provides an estimate of digital and analog information stored over the past decades.[3] This report noted that digital information surpassed analog information around 2003. At that time, there were about 10 million terabytes of digital information, which is roughly about 10 million storage drives today. However, a more relevant finding from this report was that our footprint of digital information is growing at exponential rates. Figure 1-1 shows the findings of this report; notice that every other year, the world's information has grown tenfold.

With the ambition to provide tools capable of searching all of this new digital information, many companies attempted to provide such functionality with what we know today as search engines, used when searching the web. Given the vast amount of digital information, managing information at this scale was a challenging problem. Search engines were unable to store all of the web page information required to support web searches in a single computer. This meant that they had to split information into several files and store them across many machines. This approach became known as the *Google File System*, and was presented in a research paper published in 2003 by Google.[4]

---

1 Laudon KC, Traver CG, Laudon JP (1996). "Information technology and systems." *Cambridge, MA: Course Technology.*

2 Ceruzzi PE (2012). *Computing: a concise history.* MIT Press.

3 Group WB (2016). *The Data Revolution.* World Bank Publications.

4 Ghemawat S, Gobioff H, Leung S (2003). "The Google File System." In *Proceedings of the Nineteenth ACM Symposium on Operating Systems Principles.* ISBN 1-58113-757-5.

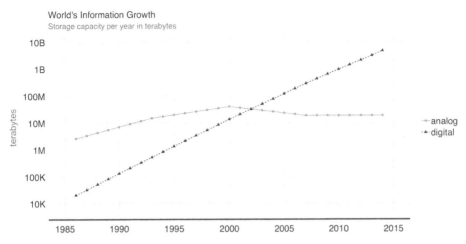

Figure 1-1. World's capacity to store information

Hadoop

One year later, Google published a new paper describing how to perform operations across the Google File System, an approach that came to be known as *MapReduce*.[5] As you would expect, there are two operations in MapReduce: map and reduce. The *map operation* provides an arbitrary way to transform each file into a new file, whereas the *reduce operation* combines two files. Both operations require custom computer code, but the MapReduce framework takes care of automatically executing them across many computers at once. These two operations are sufficient to process all the data available on the web, while also providing enough flexibility to extract meaningful information from it.

For example, as illustrated in Figure 1-2, we can use MapReduce to count words in two different text files stored in different machines. The map operation splits each word in the original file and outputs a new word-counting file with a mapping of words and counts. The reduce operation can be defined to take two word-counting files and combine them by aggregating the totals for each word; this last file will contain a list of word counts across all the original files.

Counting words is often the most basic MapReduce example, but we can also use MapReduce for much more sophisticated and interesting applications. For instance, we can use it to rank web pages in Google's *PageRank* algorithm, which assigns ranks

---

5 Dean J, Ghemawat S (2004). "MapReduce: Simplified data processing on large clusters." In *USENIX Symposium on Operating System Design and Implementation (OSDI)*.

to web pages based on the count of hyperlinks linking to a web page and the rank of the page linking to it.

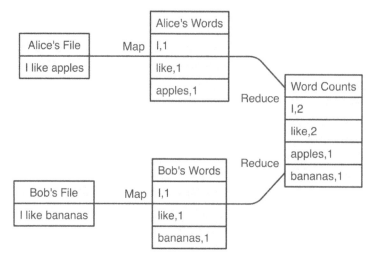

Figure 1-2. MapReduce example counting words across files

After these papers were released by Google, a team at Yahoo worked on implementing the Google File System and MapReduce as a single open source project. This project was released in 2006 as *Hadoop*, with the Google File System implemented as the *Hadoop Distributed File System* (HDFS). The Hadoop project made distributed file-based computing accessible to a wider range of users and organizations, making MapReduce useful beyond web data processing.

Although Hadoop provided support to perform MapReduce operations over a distributed file system, it still required MapReduce operations to be written with code every time a data analysis was run. To improve upon this tedious process, the *Hive* project, released in 2008 by Facebook, brought *Structured Query Language* (SQL) support to Hadoop. This meant that data analysis could now be performed at large scale without the need to write code for each MapReduce operation; instead, one could write generic data analysis statements in SQL, which are much easier to understand and write.

Spark

In 2009, *Apache Spark* began as a research project at UC Berkeley's AMPLab to improve on MapReduce. Specifically, Spark provided a richer set of verbs beyond MapReduce to facilitate optimizing code running in multiple machines. Spark also loaded data in-memory, making operations much faster than Hadoop's on-disk storage. One of the earliest results showed that running *logistic regression*, a data modeling technique that we will introduce in Chapter 4, allowed Spark to run 10 times faster than Hadoop by making use of in-memory datasets.[6] A chart similar to Figure 1-3 was presented in the original research publication.

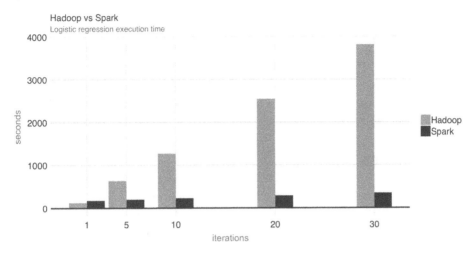

Figure 1-3. Logistic regression performance in Hadoop and Spark

Even though Spark is well known for its in-memory performance, it was designed to be a general execution engine that works both in-memory and on-disk. For instance, Spark has set new records in large-scale sorting (*http://sortbenchmark.org/*), for which data was not loaded in-memory; rather, Spark made improvements in network serialization, network shuffling, and efficient use of the CPU's cache to dramatically enhance performance. If you needed to sort large amounts of data, there was no other system in the world faster than Spark.

To give you a sense of how much faster and efficient Spark is, it takes 72 minutes and 2,100 computers to sort 100 terabytes of data using Hadoop, but only 23 minutes and 206 computers using Spark (*https://oreil.ly/Duram*). In addition, Spark holds the cloud sorting record (*https://oreil.ly/lHMjg*), which makes it the most cost-effective solution for sorting large-datasets in the cloud.

	Hadoop record	Spark record
Data size	102.5 TB	100 TB
Elapsed time	72 mins	23 mins
Nodes	2,100	206

---

6 Zaharia M, Chowdhury M, Franklin MJ, Shenker S, Stoica I (2010). "Spark: Cluster computing with working sets." *HotCloud*, 10(10-10), 95.

	Hadoop record	Spark record
Cores	50,400	6,592
Disk	3,150 GB/s	618 GB/s
Network	10 GB/s	10 GB/s
Sort rate	1.42 TB/min	4.27 TB/min
Sort rate/node	0.67 GB/min	20.7 GB/min

Spark is also easier to use than Hadoop; for instance, the word-counting MapReduce example takes about 50 lines of code in Hadoop, but it takes only 2 lines of code in Spark. As you can see, Spark is much faster, more efficient, and easier to use than Hadoop.

In 2010, Spark was released as an open source project and then donated to the Apache Software Foundation in 2013. Spark is licensed under Apache 2.0 (*https:// oreil.ly/cNH5p*), which allows you to freely use, modify, and distribute it. Spark then reached more than 1,000 contributors, making it one of the most active projects in the Apache Software Foundation.

This gives an overview of how Spark came to be, which we can now use to formally introduce Apache Spark as defined on the project's website (*http://spark.apache.org*):

> Apache Spark is a unified analytics engine for large-scale data processing.

To help us understand this definition of Apache Spark, we break it down as follows:

Unified
Spark supports many libraries, cluster technologies, and storage systems.

Analytics
Analytics is the discovery and interpretation of data to produce and communicate information.

Engine
Spark is expected to be efficient and generic.

Large-scale
You can interpret large-scale as *cluster*-scale, a set of connected computers working together.

Spark is described as an *engine* because it's generic and efficient. It's generic because it optimizes and executes generic code; that is, there are no restrictions as to what type of code you can write in Spark. It is efficient, because, as we mentioned earlier, Spark much faster than other technologies by making efficient use of memory, network, and CPUs to speed data processing algorithms in computing clusters.

This makes Spark ideal in many *analytics* projects like ranking movies at Netflix (*http://spark.apache.org*), aligning protein sequences (*https://bit.ly/2KUZEdb*), or analyzing high-energy physics at CERN (*https://bit.ly/2KoTGlc*).

As a *unified* platform, Spark is expected to support many cluster technologies and multiple data sources, which you learn about in Chapter 6 and Chapter 8, respectively. It is also expected to support many different libraries like Spark SQL, MLlib, GraphX, and Spark Streaming; libraries that you can use for analysis, modeling, graph processing, and real-time data processing, respectively. In summary, Spark is a platform providing access to clusters, data sources, and libraries for large-scale computing, as illustrated in Figure 1-4.

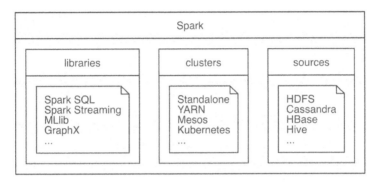

Figure 1-4. Spark as a unified analytics engine for large-scale data processing

Describing Spark as *large scale* implies that a good use case for Spark is tackling problems that can be solved with multiple machines. For instance, when data does not fit on a single disk drive or into memory, Spark is a good candidate to consider. However, you can also consider it for problems that might not be large scale, but for which using multiple computers could speed up computation. For instance, CPU-intensive models and scientific simulations also benefit from running in Spark.

Therefore, Spark is good at tackling large-scale data-processing problems, usually known as *big data* (*https://bit.ly/2XnLHec*) (datasets that are more voluminous and complex than traditional ones) but it is also good at tackling large-scale computation problems, known as *big compute* (*http://bit.ly/2OVzHOc*) (tools and approaches using a large amount of CPU and memory resources in a coordinated way). Big data often requires big compute, but big compute does not necessarily require big data (*https://bit.ly/2FhjStV*).

Big data and big compute problems are usually easy to spot—if the data does not fit into a single machine, you might have a big data problem; if the data fits into a single machine but processing it takes days, weeks, or even months to compute, you might have a big compute problem.

However, there is also a third problem space for which neither data nor compute is necessarily large scale and yet there are significant benefits to using cluster computing frameworks like Spark. For this third problem space, there are a few use cases:

Velocity

Suppose that you have a dataset of 10 GB and a process that takes 30 minutes to run over this data—this is neither big compute nor big data by any means. However, if you happen to be researching ways to improve the accuracy of your models, reducing the runtime down to three minutes is a significant improvement, which can lead to meaningful advances and productivity gains by increasing the velocity at which you can analyze data. Alternatively, you might need to process data faster—for stock trading, for instance. Whereas three minutes could seem fast enough, it can be far too slow for real-time data processing, for which you might need to process data in a few seconds—or even a few milliseconds.

Variety

You could also have an efficient process to collect data from many sources into a single location, usually a database; this process could be already running efficiently and close to real time. Such processes are known as *Extract, Transform, Load* (ETL); data is extracted from multiple sources, transformed to the required format, and loaded into a single data store. Although this has worked for years, the trade-off from this approach is that adding a new data source is expensive. Because the system is centralized and tightly controlled, making changes could cause the entire process to halt; therefore, adding new data source usually takes too long to be implemented. Instead, you can store all data in its natural format and process it as needed using cluster computing, an architecture known as a *data lake*. In addition, storing data in its raw format allows you to process a variety of new file formats like images, audio, and video without having to figure out how to fit them into conventional structured storage systems.

Veracity

When using many data sources, you might find the data quality varies greatly between them, which requires special analysis methods to improve their accuracy. For instance, suppose that you have a table of cities with values like San Francisco, Seattle, and Boston. What happens when data contains a misspelled entry like "Bston"? In a relational database, this invalid entry might be dropped. However, dropping values is not necessarily the best approach in all cases; you might want to correct this field by making use of geocodes, cross-referencing data sources, or attempting a best-effort match. Therefore, understanding and improving the veracity of the original data source can lead to more accurate results.

If we include "volume" as a synonym for big data, you get the mnemonic people refer to as the four Vs of big data (*http://bit.ly/2MkF1sp*); others have expanded this to five (*https://oreil.ly/gU9rP*) or even 10 Vs of big data (*http://bit.ly/2KBOGbM*). Mnemonics aside, cluster computing is being used today in more innovative ways, and is not uncommon to see organizations experimenting with new workflows and a variety of tasks that were traditionally uncommon for cluster computing. Much of the hype attributed to big data falls into this space where, strictly speaking, you're not handling big data, but there are still benefits from using tools designed for big data and big compute.

Our hope is that this book will help you to understand the opportunities and limitations of cluster computing and, specifically, the opportunities and limitations of using Apache Spark with R.

R

The R computing language has its origins in the S language, which was created at Bell Laboratories. Rick Becker explained in useR 2016 (*https://bit.ly/2MSTm0j*) that at that time in Bell Labs, computing was done by calling subroutines written in the Fortran language, which, apparently, were not pleasant to deal with. The S computing language was designed as an interface language to solve particular problems without having to worry about other languages, such as Fortran. The creator of S, John Chambers (*http://bit.ly/2Z5QygX*), shows in Figure 1-5 how S was designed to provide an interface that simplifies data processing; his co-creator presented this during *useR! 2016* as the original diagram that inspired the creation of S.

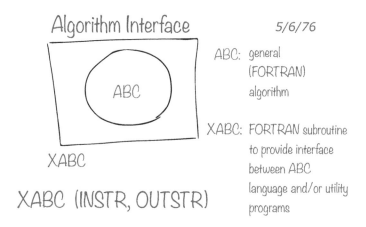

Figure 1-5. Interface language diagram by John Chambers (Rick Becker useR 2016)

R is a modern and free implementation of S. Specifically, according to the R Project for Statistical Computing (*https://www.r-project.org*):

R is a programming language and free software environment for statistical computing and graphics.

While working with data, we believe there are two strong arguments for using R:

R language

R was designed by statisticians for statisticians, meaning that this is one of the few successful languages designed for nonprogrammers, so learning R will probably feel more natural. Additionally, because the R language was designed to be an interface to other tools and languages, R allows you to focus more on understanding data and less on the particulars of computer science and engineering.

R community

The R community provides a rich package archive provided by the Comprehensive R Archive Network (*https://cran.r-project.org/*) (CRAN), which allows you to install ready-to-use packages to perform many tasks—most notably high-quality data manipulation, visualization, and statistical models, many of which are available only in R. In addition, the R community is a welcoming and active group of talented individuals motivated to help you succeed. Many packages provided by the R community make R, by far, the best option for statistical computing. Some of the most downloaded R packages include: dplyr (*http://bit.ly/2YS3PP3*) to manipulate data, cluster (*http://bit.ly/307Tuv7*) to analyze clusters, and ggplot2 (*http://bit.ly/2ZaWMjY*) to visualize data. Figure 1-6 quantifies the growth of the R community by plotting daily downloads of R packages in CRAN.

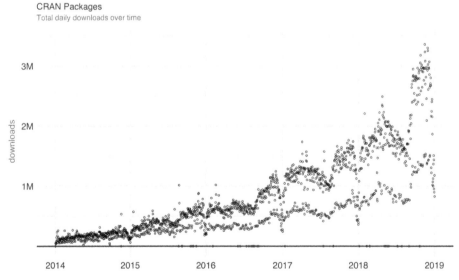

Figure 1-6. Daily downloads of CRAN packages

Aside from statistics, R is also used in many other fields. The following areas are particularly relevant to this book:

Data science

Data science is based on knowledge and practices from statistics and computer science that turn raw data into understanding[7] by using data analysis and modeling techniques. Statistical methods provide a solid foundation to understand the world and perform predictions, while the automation provided by computing methods allows us to simplify statistical analysis and make it much more accessible. Some have advocated that statistics should be renamed data science;[8] however, data science goes beyond statistics by also incorporating advances in computing.[9] This book presents analysis and modeling techniques common in statistics but applied to large datasets, which requires incorporating advances in distributed computing.

Machine learning

Machine learning uses practices from statistics and computer science; however, it is heavily focused on automation and prediction. For instance, Arthur Samuel coined the term *machine learning* while automating a computer program to play checkers.[10] Although we could perform data science on particular games, writing a program to play checkers requires us to automate the entire process. Therefore, this falls in the realm of machine learning, not data science. Machine learning makes it possible for many users to take advantage of statistical methods without being aware of using them. One of the first important applications of machine learning was to filter spam emails. In this case, it's just not feasible to perform data analysis and modeling over each email account; therefore, machine learning automates the entire process of finding spam and filtering it out without having to involve users at all. This book presents the methods to transition data science workflows into fully automated machine learning methods—for instance, by providing support to build and export Spark pipelines that can be easily reused in automated environments.

Deep learning

Deep learning builds on knowledge of statistics, data science, and machine learning to define models loosely inspired by biological nervous systems. Deep learn-

---

7 Wickham H, Grolemund G (2016). *R for data science: import, tidy, transform, visualize, and model data.* O'Reilly Media, Inc.

8 Wu CJ (1997). "Statistics = Data Science?"

9 Cleveland WS (2001). "Data Science: An Action Plan for Expanding the Technical Areas of the Field of Statistics?"

10 Samuel AL (1959). "Some studies in machine learning using the game of checkers." *IBM Journal of research and development*, 3(3), 210–229.

ing models evolved from neural network models after the vanishing-gradient problem was resolved by training one layer at a time,[11] and have proven useful in image and speech recognition tasks. For instance, in voice assistants like Siri, Alexa, Cortana, or Google Assistant, the model performing the audio-to-text conversion is most likely based on deep learning models. Although Graphic Processing Units (GPUs) have been successfully used to train deep learning models,[12] some datasets cannot be processed in a single GPU. It is also the case that deep learning models require huge amounts of data, which needs to be preprocessed across many machines before it can be fed into a single GPU for training. This book doesn't make any direct references to deep learning models; however, you can use the methods we present in this book to prepare data for deep learning and, in the years to come, using deep learning with large-scale computing will become a common practice. In fact, recent versions of Spark have already introduced execution models optimized for training deep learning in Spark.

When working in any of the previous fields, you will be faced with increasingly large datasets or increasingly complex computations that are slow to execute or at times even impossible to process in a single computer. However, it is important to understand that Spark does not need to be the answer to all our computations problems; instead, when faced with computing challenges in R, using the following techniques can be as effective:

Sampling

A first approach to try is to reduce the amount of data being handled, through sampling. However, we must sample the data properly by applying sound statistical principles. For instance, selecting the top results is not sufficient in sorted datasets; with simple random sampling, there might be underrepresented groups, which we could overcome with stratified sampling, which in turn adds complexity to properly select categories. It's beyond the scope of this book to teach how to properly perform statistical sampling, but many resources are available on this topic.

Profiling

You can try to understand why a computation is slow and make the necessary improvements. A profiler is a tool capable of inspecting code execution to help identify bottlenecks. In R, the R profiler, the `profvis` (*http://bit.ly/2OXGabw*) R package, and RStudio profiler feature (*https://bit.ly/2RqJPw8*) allow you to easily to retrieve and visualize a profile; however, it's not always trivial to optimize.

---

11 Hinton GE, Osindero S, Teh Y (2006). "A fast learning algorithm for deep belief nets." *Neural computation*, 18(7), 1527–1554.

12 Krizhevsky A, Sutskever I, Hinton GE (2012). "Imagenet classification with deep convolutional neural networks." In *Advances in neural information processing systems*, 1097–1105.

Scaling up

Speeding up computation is usually possible by buying faster or more capable hardware (say, increasing your machine memory, upgrading your hard drive, or procuring a machine with many more CPUs); this approach is known as *scaling up*. However, there are usually hard limits as to how much a single computer can scale up, and even with significant CPUs, you need to find frameworks that parallelize computation efficiently.

Scaling out

Finally, we can consider spreading computation and storage across multiple machines. This approach provides the highest degree of scalability because you can potentially use an arbitrary number of machines to perform a computation. This approach is commonly known as *scaling out*. However, spreading computation effectively across many machines is a complex endeavor, especially without using specialized tools and frameworks like Apache Spark.

This last point brings us closer to the purpose of this book, which is to bring the power of distributed computing systems provided by Apache Spark to solve meaningful computation problems in data science and related fields, using R.

sparklyr

When you think of the computation power that Spark provides and the ease of use of the R language, it is natural to want them to work together, seamlessly. This is also what the R community expected: an R package that would provide an interface to Spark that was easy to use, compatible with other R packages, and available in CRAN. With this goal, we started developing sparklyr. The first version, sparklyr 0.4 (*http://bit.ly/2Zhgevy*), was released during the *useR! 2016* conference. This first version included support for dplyr, DBI, modeling with MLlib, and an extensible API that enabled extensions like H2O (*https://www.h2o.ai/*)'s rsparkling (*http://bit.ly/2z348qO*) package. Since then, many new features and improvements have been made available through sparklyr 0.5 (*http://bit.ly/sparklyr05*), 0.6 (*http://bit.ly/sparklyr06*), 0.7 (*http://bit.ly/sparklyr07*), 0.8 (*http://bit.ly/sparklyr08*), 0.9 (*http://bit.ly/2TBnKMt*) and 1.0 (*http://bit.ly/sparklyr10*).

Officially, sparklyr is an R interface for Apache Spark. It's available in CRAN and works like any other CRAN package, meaning that it's agnostic to Spark versions, it's easy to install, it serves the R community, it embraces other packages and practices from the R community, and so on. It's hosted in GitHub (*http://bit.ly/30b5NGT*) and licensed under Apache 2.0, which allows you to clone, modify, and contribute back to this project.

When thinking of who should use sparklyr, the following roles come to mind:

New users
> For new users, it is our belief that sparklyr provides the easiest way to get started with Spark. Our hope is that the early chapters of this book will get you up and running with ease and set you up for long-term success.

Data scientists
> For data scientists who already use and love R, sparklyr integrates with many other R practices and packages like dplyr, magrittr, broom, DBI, tibble, rlang, and many others, which will make you feel at home while working with Spark. For those new to R and Spark, the combination of high-level workflows available in sparklyr and low-level extensibility mechanisms make it a productive environment to match the needs and skills of every data scientist.

Expert users
> For those users who are already immersed in Spark and can write code natively in Scala, consider making your Spark libraries available as an R package to the R community, a diverse and skilled community that can put your contributions to good use while moving open science (*http://bit.ly/2yZLrVd*) forward.

We wrote this book to describe and teach the exciting overlap between Apache Spark and R. sparklyr is the R package that brings together these communities, expectations, future directions, packages, and package extensions. We believe that there is an opportunity to use this book to bridge the R and Spark communities: to present to the R community why Spark is exciting, and to the Spark community what makes R great. Both communities are solving very similar problems with a set of different skills and backgrounds; therefore, it is our hope that sparklyr can be a fertile ground for innovation, a welcoming place for newcomers, a productive environment for experienced data scientists, and an open community where cluster computing, data science, and machine learning can come together.

Recap

This chapter presented Spark as a modern and powerful computing platform, R as an easy-to-use computing language with solid foundations in statistical methods, and sparklyr as a project bridging both technologies and communities. In a world in which the total amount of information is growing exponentially, learning how to analyze data at scale will help you to tackle the problems and opportunities humanity is facing today. However, before we start analyzing data, Chapter 2 will equip you with the tools you will need throughout the rest of this book. Be sure to follow each step carefully and take the time to install the recommended tools, which we hope will become familiar resources that you use and love.

## CHAPTER 2

Getting Started

I always wanted to be a wizard.

—Samwell Tarly

After reading Chapter 1, you should now be familiar with the kinds of problems that Spark can help you solve. And it should be clear that Spark solves problems by making use of multiple computers when data does not fit in a single machine or when computation is too slow. If you are newer to R, it should also be clear that combining Spark with data science tools like `ggplot2` for visualization and `dplyr` to perform data transformations brings a promising landscape for doing data science at scale. We also hope you are excited to become proficient in large-scale computing.

In this chapter, we take a tour of the tools you'll need to become proficient in Spark. We encourage you to walk through the code in this chapter because it will force you to go through the motions of analyzing, modeling, reading, and writing data. In other words, you will need to do some wax-on, wax-off, repeat before you get fully immersed in the world of Spark.

In Chapter 3 we dive into analysis followed by modeling, which presents examples using a single-cluster machine: your personal computer. Subsequent chapters introduce cluster computing and the concepts and techniques that you'll need to successfully run code across multiple machines.

Overview

From R, getting started with Spark using `sparklyr` and a local cluster is as easy as installing and loading the `sparklyr` package followed by installing Spark using `sparklyr`; however, we assume you are starting with a brand new computer running Windows, macOS, or Linux, so we'll walk you through the prerequisites before connecting to a local Spark cluster.

Although this chapter is designed to help you get ready to use Spark on your personal computer, it's also likely that some readers will already have a Spark cluster available or might prefer to get started with an online Spark cluster. For instance, Databricks hosts a free community edition (*http://bit.ly/31MfKuV*) of Spark that you can easily access from your web browser. If you end up choosing this path, skip to "Prerequisites" on page 16, but make sure you consult the proper resources for your existing or online Spark cluster.

Either way, after you are done with the prerequisites, you will first learn how to connect to Spark. We then present the most important tools and operations that you'll use throughout the rest of this book. Less emphasis is placed on teaching concepts or how to use them—we can't possibly explain modeling or streaming in a single chapter. However, going through this chapter should give you a brief glimpse of what to expect and give you the confidence that you have the tools correctly configured to tackle more challenging problems later on.

The tools you'll use are mostly divided into R code and the Spark web interface. All Spark operations are run from R; however, monitoring execution of distributed operations is performed from Spark's web interface, which you can load from any web browser. We then disconnect from this local cluster, which is easy to forget to do but highly recommended while working with local clusters—and in shared Spark clusters as well!

We close this chapter by walking you through some of the features that make using Spark with RStudio easier; more specifically, we present the RStudio extensions that sparklyr implements. However, if you are inclined to use Jupyter Notebooks or if your cluster is already equipped with a different R user interface, rest assured that you can use Spark with R through plain R code. Let's move along and get your prerequisites properly configured.

Prerequisites

R can run in many platforms and environments; therefore, whether you use Windows, Mac, or Linux, the first step is to install R from r-project.org (*https://r-project.org/*); detailed instructions are provided in "Installing R" on page 251.

Most people use programming languages with tools to make them more productive; for R, RStudio is such a tool. Strictly speaking, RStudio is an *integrated development environment* (IDE), which also happens to support many platforms and environments. We strongly recommend you install RStudio if you haven't done so already; see details in "Installing RStudio" on page 253.

When using Windows, we recommend avoiding directories with spaces in their path. If running getwd() from R returns a path with spaces, consider switching to a path with no spaces using setwd("path") or by creating an RStudio project in a path with no spaces.

Additionally, because Spark is built in the Scala programming language, which is run by the Java Virtual Machine (JVM), you also need to install Java 8 on your system. It is likely that your system already has Java installed, but you should still check the version and update or downgrade as described in "Installing Java" on page 252. You can use the following R command to check which version is installed on your system:

```
system("java -version")

java version "1.8.0_201"
Java(TM) SE Runtime Environment (build 1.8.0_201-b09)
Java HotSpot(TM) 64-Bit Server VM (build 25.201-b09, mixed mode)
```

You can also use the JAVA_HOME environment variable to point to a specific Java version by running Sys.setenv(JAVA_HOME = "path-to-java-8"); either way, before moving on to installing sparklyr, make sure that Java 8 is the version available for R.

Installing sparklyr

As with many other R packages, you can install sparklyr from CRAN (*http://bit.ly/2KLyaoE*) as follows:

```
install.packages("sparklyr")
```

The examples in this book assume you are using the latest version of sparklyr. You can verify your version is as new as the one we are using by running the following:

```
packageVersion("sparklyr")
[1] '1.0.2'
```

Installing Spark

Start by loading sparklyr:

```
library(sparklyr)
```

This makes all sparklyr functions available in R, which is really helpful; otherwise, you would need to run each sparklyr command prefixed with sparklyr::.

You can easily install Spark by running spark_install(). This downloads, installs, and configures the latest version of Spark locally on your computer; however, because we've written this book with Spark 2.3, you should also install this version to make sure that you can follow all the examples provided without any surprises:

```
spark_install("2.3")
```

You can display all of the versions of Spark that are available for installation by running the following:

```
spark_available_versions()

##    spark
## 1    1.6
## 2    2.0
## 3    2.1
## 4    2.2
## 5    2.3
## 6    2.4
```

You can install a specific version by using the Spark version and, optionally, by also specifying the Hadoop version. For instance, to install Spark 1.6.3, you would run:

```
spark_install(version = "1.6.3")
```

You can also check which versions are installed by running this command:

```
spark_installed_versions()

  spark hadoop                              dir
7 2.3.1    2.7 /spark/spark-2.3.1-bin-hadoop2.7
```

The path where Spark is installed is known as Spark's *home*, which is defined in R code and system configuration settings with the SPARK_HOME identifier. When you are using a local Spark cluster installed with sparklyr, this path is already known and no additional configuration needs to take place.

Finally, to uninstall a specific version of Spark you can run spark_uninstall() by specifying the Spark and Hadoop versions, like so:

```
spark_uninstall(version = "1.6.3", hadoop = "2.6")
```

The default installation paths are *~/spark* for macOS and Linux, and *%LOCALAPPDATA%/spark* for Windows. To customize the installation path, you can run options(spark.install.dir = "installation-path") before spark_install() and spark_con nect().

Connecting

It's important to mention that, so far, we've installed only a local Spark cluster. A local cluster is really helpful to get started, test code, and troubleshoot with ease. Later chapters explain where to find, install, and connect to real Spark clusters with many machines, but for the first few chapters, we focus on using local clusters.

To connect to this local cluster, simply run the following:

```
library(sparklyr)
sc <- spark_connect(master = "local", version = "2.3")
```

 If you are using your own or online Spark cluster, make sure that you connect as specified by your cluster administrator or the online documentation. If you need some pointers, you can take a quick look at Chapter 7, which explains in detail how to connect to any Spark cluster.

The master parameter identifies which is the "main" machine from the Spark cluster; this machine is often called the *driver node*. While working with real clusters using many machines, you'll find that most machines will be worker machines and one will be the master. Since we have only a local cluster with just one machine, we will default to using "local" for now.

After a connection is established, spark_connect() retrieves an active Spark connection, which most code usually names sc; you will then make use of sc to execute Spark commands.

If the connection fails, Chapter 7 contains a troubleshooting section that can help you to resolve your connection issue.

Using Spark

Now that you are connected, we can run a few simple commands. For instance, let's start by copying the mtcars dataset into Apache Spark by using copy_to():

```
cars <- copy_to(sc, mtcars)
```

The data was copied into Spark, but we can access it from R using the cars reference. To print its contents, we can simply type *cars*:

```
cars
```

```
# Source: spark<mtcars> [?? x 11]
      mpg   cyl  disp    hp  drat    wt  qsec    vs    am  gear  carb
    <dbl> <dbl> <dbl> <dbl> <dbl> <dbl> <dbl> <dbl> <dbl> <dbl> <dbl>
 1  21       6   160   110  3.9   2.62  16.5     0     1     4     4
 2  21       6   160   110  3.9   2.88  17.0     0     1     4     4
 3  22.8     4   108    93  3.85  2.32  18.6     1     1     4     1
 4  21.4     6   258   110  3.08  3.22  19.4     1     0     3     1
 5  18.7     8   360   175  3.15  3.44  17.0     0     0     3     2
 6  18.1     6   225   105  2.76  3.46  20.2     1     0     3     1
 7  14.3     8   360   245  3.21  3.57  15.8     0     0     3     4
 8  24.4     4   147.   62  3.69  3.19  20       1     0     4     2
 9  22.8     4   141.   95  3.92  3.15  22.9     1     0     4     2
10  19.2     6   168.  123  3.92  3.44  18.3     1     0     4     4
# ... with more rows
```

Congrats! You have successfully connected and loaded your first dataset into Spark.

Let's explain what's going on in copy_to(). The first parameter, sc, gives the function a reference to the active Spark connection that was created earlier with spark_con nect(). The second parameter specifies a dataset to load into Spark. Now, copy_to() returns a reference to the dataset in Spark, which R automatically prints. Whenever a Spark dataset is printed, Spark *collects* some of the records and displays them for you. In this particular case, that dataset contains only a few rows describing automobile models and some of their specifications like horsepower and expected miles per gallon.

Web Interface

Most of the Spark commands are executed from the R console; however, monitoring and analyzing execution is done through Spark's web interface, shown in Figure 2-1. This interface is a web application provided by Spark that you can access by running:

```
spark_web(sc)
```

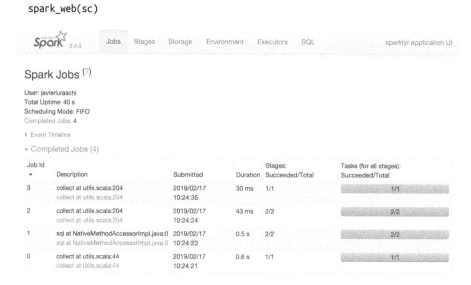

Figure 2-1. The Apache Spark web interface

Printing the cars dataset collected a few records to be displayed in the R console. You can see in the Spark web interface that a job was started to collect this information back from Spark. You can also select the Storage tab to see the mtcars dataset cached in memory in Spark, as shown in Figure 2-2.

Notice that this dataset is fully loaded into memory, as indicated by the Fraction Cached column, which shows 100%; thus, you can see exactly how much memory this dataset is using through the Size in Memory column.

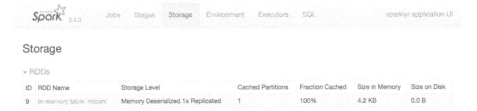

Spark 2.4.0   Jobs   Stages   Storage   Environment   Executors   SQL   sparklyr application UI

Storage

▾ RDDs

ID	RDD Name	Storage Level	Cached Partitions	Fraction Cached	Size in Memory	Size on Disk
9	In-memory table 'mtcars'	Memory Deserialized 1x Replicated	1	100%	4.2 KB	0.0 B

Figure 2-2. The Storage tab on the Apache Spark web interface

The Executors tab, shown in Figure 2-3, provides a view of your cluster resources. For local connections, you will find only one executor active with only 2 GB of memory allocated to Spark, and 384 MB available for computation. In Chapter 9 you will learn how to request more compute instances and resources, and how memory is allocated.

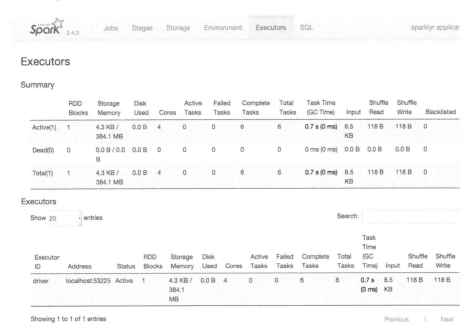

Spark 2.4.0   Jobs   Stages   Storage   Environment   Executors   SQL   sparklyr applica

Executors

Summary

	RDD Blocks	Storage Memory	Disk Used	Cores	Active Tasks	Failed Tasks	Complete Tasks	Total Tasks	Task Time (GC Time)	Input	Shuffle Read	Shuffle Write	Blacklisted
Active(1)	1	4.3 KB / 384.1 MB	0.0 B	4	0	0	6	6	0.7 s (0 ms)	8.5 KB	118 B	118 B	0
Dead(0)	0	0.0 B / 0.0 B	0.0 B	0	0	0	0	0	0 ms (0 ms)	0.0 B	0.0 B	0.0 B	0
Total(1)	1	4.3 KB / 384.1 MB	0.0 B	4	0	0	6	6	0.7 s (0 ms)	8.5 KB	118 B	118 B	0

Executors

Show 20 entries   Search:

Executor ID	Address	Status	RDD Blocks	Storage Memory	Disk Used	Cores	Active Tasks	Failed Tasks	Complete Tasks	Total Tasks	Task Time (GC Time)	Input	Shuffle Read	Shuffle Write
driver	localhost:53225	Active	1	4.3 KB / 384.1 MB	0.0 B	4	0	0	6	6	0.7 s (0 ms)	8.5 KB	118 B	118 B

Showing 1 to 1 of 1 entries   Previous   1   Next

Figure 2-3. The Executors tab on the Apache Spark web interface

The last tab to explore is the Environment tab, shown in Figure 2-4; this tab lists all of the settings for this Spark application, which we look at in Chapter 9. As you will learn, most settings don't need to be configured explicitly, but to properly run them at scale, you need to become familiar with some of them, eventually.

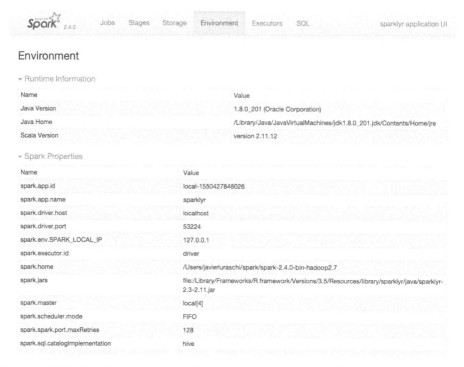

Environment

▾ Runtime Information

Name	Value
Java Version	1.8.0_201 (Oracle Corporation)
Java Home	/Library/Java/JavaVirtualMachines/jdk1.8.0_201.jdk/Contents/Home/jre
Scala Version	version 2.11.12

▾ Spark Properties

Name	Value
spark.app.id	local-1550427846026
spark.app.name	sparklyr
spark.driver.host	localhost
spark.driver.port	53224
spark.env.SPARK_LOCAL_IP	127.0.0.1
spark.executor.id	driver
spark.home	/Users/javierluraschi/spark/spark-2.4.0-bin-hadoop2.7
spark.jars	file:/Library/Frameworks/R.framework/Versions/3.5/Resources/library/sparklyr/java/sparklyr-2.3-2.11.jar
spark.master	local[4]
spark.scheduler.mode	FIFO
spark.spark.port.maxRetries	128
spark.sql.catalogImplementation	hive

Figure 2-4. The Environment tab on the Apache Spark web interface

Next, you will make use of a small subset of the practices that we cover in depth in Chapter 3.

Analysis

When using Spark from R to analyze data, you can use SQL (Structured Query Language) or dplyr (a grammar of data manipulation). You can use SQL through the DBI package; for instance, to count how many records are available in our cars dataset, we can run the following:

```
library(DBI)
dbGetQuery(sc, "SELECT count(*) FROM mtcars")

  count(1)
1       32
```

When using dplyr, you write less code, and it's often much easier to write than SQL. This is precisely why we won't make use of SQL in this book; however, if you are proficient in SQL, this is a viable option for you. For instance, counting records in dplyr is more compact and easier to understand:

```
library(dplyr)
count(cars)

# Source: spark<?> [?? x 1]
      n
  <dbl>
1    32
```

In general, we usually start by analyzing data in Spark with dplyr, followed by sampling rows and selecting a subset of the available columns. The last step is to collect data from Spark to perform further data processing in R, like data visualization. Let's perform a very simple data analysis example by selecting, sampling, and plotting the cars dataset in Spark:

```
select(cars, hp, mpg) %>%
  sample_n(100) %>%
  collect() %>%
  plot()
```

The plot in Figure 2-5 shows that as we increase the horsepower in a vehicle, its fuel efficiency measured in miles per gallon decreases. Although this is insightful, it's difficult to predict numerically how increased horsepower would affect fuel efficiency. Modeling can help us overcome this.

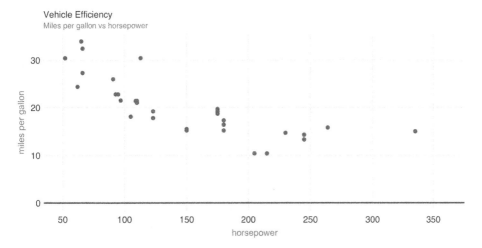

Figure 2-5. Horsepower versus miles per gallon

Modeling

Although data analysis can take you quite far toward understanding data, building a mathematical model that describes and generalizes the dataset is quite powerful. In Chapter 1 you learned that the fields of machine learning and data science make use of mathematical models to perform predictions and find additional insights.

For instance, we can use a linear model to approximate the relationship between fuel efficiency and horsepower:

```
model <- ml_linear_regression(cars, mpg ~ hp)
model

Formula: mpg ~ hp

Coefficients:
(Intercept)           hp
 30.09886054 -0.06822828
```

Now we can use this model to predict values that are not in the original dataset. For instance, we can add entries for cars with horsepower beyond 250 and also visualize the predicted values, as shown in Figure 2-6.

```
model %>%
  ml_predict(copy_to(sc, data.frame(hp = 250 + 10 * 1:10))) %>%
  transmute(hp = hp, mpg = prediction) %>%
  full_join(select(cars, hp, mpg)) %>%
  collect() %>%
  plot()
```

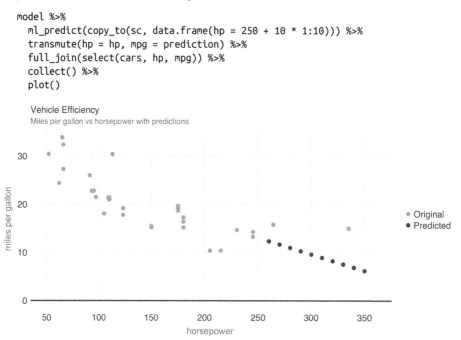

Figure 2-6. Horsepower versus miles per gallon with predictions

Even though the previous example lacks many of the appropriate techniques that you should use while modeling, it's also a simple example to briefly introduce the modeling capabilities of Spark. We introduce all of the Spark models, techniques, and best practices in Chapter 4.

Data

For simplicity, we copied the `mtcars` dataset into Spark; however, data is usually not copied into Spark. Instead, data is read from existing data sources in a variety of formats, like plain text, CSV, JSON, Java Database Connectivity (JDBC), and many more, which we examine in detail in Chapter 8. For instance, we can export our `cars` dataset as a CSV file:

```
spark_write_csv(cars, "cars.csv")
```

In practice, we would read an existing dataset from a distributed storage system like HDFS, but we can also read back from the local file system:

```
cars <- spark_read_csv(sc, "cars.csv")
```

Extensions

In the same way that R is known for its vibrant community of package authors, at a smaller scale, many extensions for Spark and R have been written and are available to you. Chapter 10 introduces many interesting ones to perform advanced modeling, graph analysis, preprocessing of datasets for deep learning, and more.

For instance, the `sparkly.nested` extension is an R package that extends `sparklyr` to help you manage values that contain nested information. A common use case involves JSON files that contain nested lists that require preprocessing before you can do meaningful data analysis. To use this extension, we first need to install it as follows:

```
install.packages("sparklyr.nested")
```

Then, we can use the `sparklyr.nested` extension to group all of the horsepower data points over the number of cylinders:

```
sparklyr.nested::sdf_nest(cars, hp) %>%
  group_by(cyl) %>%
  summarise(data = collect_list(data))
# Source: spark<?> [?? x 2]
    cyl data
  <int> <list>
1     6 <list [7]>
2     4 <list [11]>
3     8 <list [14]>
```

Even though nesting data makes it more difficult to read, it is a requirement when you are dealing with nested data formats like JSON using the `spark_read_json()` and `spark_write_json()` functions.

Distributed R

For those few cases when a particular functionality is not available in Spark and no extension has been developed, you can consider distributing your own R code across the Spark cluster. This is a powerful tool, but it comes with additional complexity, so you should only use it as a last resort.

Suppose that we need to round all of the values across all the columns in our dataset. One approach would be running custom R code, making use of R's round() function:

```
cars %>% spark_apply(~round(.x))
# Source: spark<?> [?? x 11]
     mpg   cyl  disp    hp  drat    wt  qsec    vs    am  gear  carb
   <dbl> <dbl> <dbl> <dbl> <dbl> <dbl> <dbl> <dbl> <dbl> <dbl> <dbl>
 1    21     6   160   110     4     3    16     0     1     4     4
 2    21     6   160   110     4     3    17     0     1     4     4
 3    23     4   108    93     4     2    19     1     1     4     1
 4    21     6   258   110     3     3    19     1     0     3     1
 5    19     8   360   175     3     3    17     0     0     3     2
 6    18     6   225   105     3     3    20     1     0     3     1
 7    14     8   360   245     3     4    16     0     0     3     4
 8    24     4   147    62     4     3    20     1     0     4     2
 9    23     4   141    95     4     3    23     1     0     4     2
10    19     6   168   123     4     3    18     1     0     4     4
# … with more rows
```

If you are a proficient R user, it can be quite tempting to use spark_apply() for everything, but please, don't! spark_apply() was designed for advanced use cases where Spark falls short. You will learn how to do proper data analysis and modeling without having to distribute custom R code across your cluster.

Streaming

While processing large static datasets is the most typical use case for Spark, processing dynamic datasets in real time is also possible and, for some applications, a requirement. You can think of a streaming dataset as a static data source with new data arriving continuously, like stock market quotes. Streaming data is usually read from Kafka (an open source stream-processing software platform) or from distributed storage that receives new data continuously.

To try out streaming, let's first create an *input/* folder with some data that we will use as the input for this stream:

```
dir.create("input")
write.csv(mtcars, "input/cars_1.csv", row.names = F)
```

Then, we define a stream that processes incoming data from the *input/* folder, performs a custom transformation in R, and pushes the output into an *output/* folder:

```
stream <- stream_read_csv(sc, "input/") %>%
    select(mpg, cyl, disp) %>%
    stream_write_csv("output/")
```

As soon as the stream of real-time data starts, the *input/* folder is processed and turned into a set of new files under the *output/* folder containing the new transformed files. Since the input contained only one file, the output folder will also contain a single file resulting from applying the custom spark_apply() transformation.

```
dir("output", pattern = ".csv")

[1] "part-00000-eece04d8-7cfa-4231-b61e-f1aef8edeb97-c000.csv"
```

Up to this point, this resembles static data processing; however, we can keep adding files to the *input/* location, and Spark will parallelize and process data automatically. Let's add one more file and validate that it's automatically processed:

```
# Write more data into the stream source
write.csv(mtcars, "input/cars_2.csv", row.names = F)
```

Wait a few seconds and validate that the data is processed by the Spark stream:

```
# Check the contents of the stream destination
dir("output", pattern = ".csv")

[1] "part-00000-2d8e5c07-a2eb-449d-a535-8a19c671477d-c000.csv"
[2] "part-00000-eece04d8-7cfa-4231-b61e-f1aef8edeb97-c000.csv"
```

You should then stop the stream:

```
stream_stop(stream)
```

You can use dplyr, SQL, Spark models, or distributed R to analyze streams in real time. In Chapter 12 we properly introduce you to all the interesting transformations you can perform to analyze real-time data.

Logs

Logging is definitely less interesting than real-time data processing; however, it's a tool you should be or become familiar with. A *log* is just a text file to which Spark appends information relevant to the execution of tasks in the cluster. For local clusters, we can retrieve all the recent logs by running the following:

```
spark_log(sc)

18/10/09 19:41:46 INFO Executor: Finished task 0.0 in stage 5.0 (TID 5)...
18/10/09 19:41:46 INFO TaskSetManager: Finished task 0.0 in stage 5.0...
18/10/09 19:41:46 INFO TaskSchedulerImpl: Removed TaskSet 5.0, whose...
18/10/09 19:41:46 INFO DAGScheduler: ResultStage 5 (collect at utils...
18/10/09 19:41:46 INFO DAGScheduler: Job 3 finished: collect at utils...
```

Or, we can retrieve specific log entries containing, say, sparklyr, by using the filter parameter, as follows:

```
spark_log(sc, filter = "sparklyr")
```

```
## 18/10/09 18:53:23 INFO SparkContext: Submitted application: sparklyr
## 18/10/09 18:53:23 INFO SparkContext: Added JAR...
## 18/10/09 18:53:27 INFO Executor: Fetching spark://localhost:52930/...
## 18/10/09 18:53:27 INFO Utils: Fetching spark://localhost:52930/...
## 18/10/09 18:53:27 INFO Executor: Adding file:/private/var/folders/...
```

Most of the time, you won't need to worry about Spark logs, except in cases for which you need to troubleshoot a failed computation; in those cases, logs are an invaluable resource to be aware of. Now you know.

Disconnecting

For local clusters (really, any cluster), after you are done processing data, you should disconnect by running the following:

```
spark_disconnect(sc)
```

This terminates the connection to the cluster as well as the cluster tasks. If multiple Spark connections are active, or if the connection instance sc is no longer available, you can also disconnect all your Spark connections by running this command:

```
spark_disconnect_all()
```

Notice that exiting R, or RStudio, or restarting your R session, also causes the Spark connection to terminate, which in turn terminates the Spark cluster and cached data that is not explicitly saved.

Using RStudio

Since it's very common to use RStudio with R, sparklyr provides RStudio extensions to help simplify your workflows and increase your productivity while using Spark in RStudio. If you are not familiar with RStudio, take a quick look at "Using RStudio" on page 254. Otherwise, there are a couple extensions worth highlighting.

First, instead of starting a new connection using spark_connect() from RStudio's R console, you can use the New Connection action from the Connections tab and then select the Spark connection, which opens the dialog shown in Figure 2-7. You can then customize the versions and connect to Spark, which will simply generate the right spark_connect() command and execute this in the R console for you.

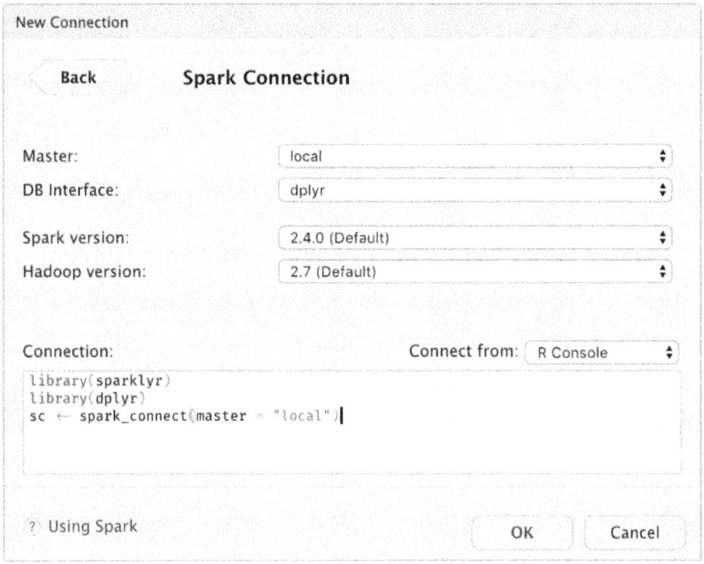

Figure 2-7. RStudio New Spark Connection dialog

After you're connected to Spark, RStudio displays your available datasets in the Connections tab, as shown in Figure 2-8. This is a useful way to track your existing datasets and provides an easy way to explore each of them.

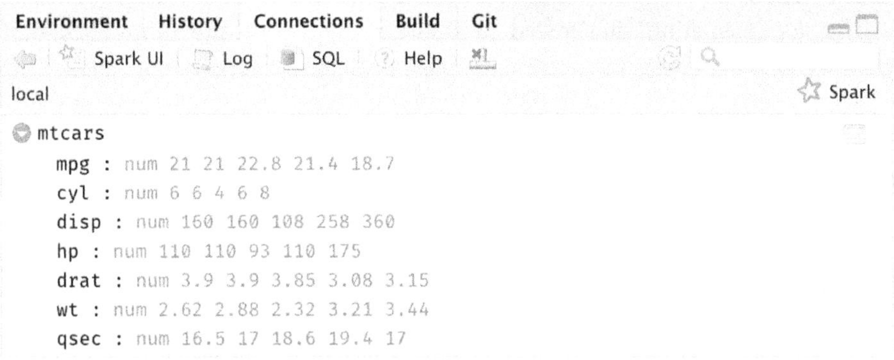

Figure 2-8. The RStudio Connections tab

Additionally, an active connection provides the following custom actions:

Spark UI

 Opens the Spark web interface; a shortcut to `spark_web(sc)`.

Log

Opens the Spark web logs; a shortcut to `spark_log(sc)`.

SQL

Opens a new SQL query. For more information about `DBI` and SQL support, see Chapter 3.

Help

Opens the reference documentation in a new web browser window.

Disconnect

Disconnects from Spark; a shortcut to `spark_disconnect(sc)`.

The rest of this book will use plain R code. It is up to you whether to execute this code in the R console, RStudio, Jupyter Notebooks, or any other tool that supports executing R code, since the examples provided in this book execute in any R environment.

Resources

While we've put significant effort into simplifying the onboarding process, there are many additional resources that can help you to troubleshoot particular issues while getting started and, in general, introduce you to the broader Spark and R communities to help you get specific answers, discuss topics, and connect with many users actively using Spark with R:

Documentation

The documentation site hosted in RStudio's Spark website (*https://spark.rstu dio.com*) should be your first stop to learn more about Spark when using R. The documentation is kept up to date with examples, reference functions, and many more relevant resources.

Blog

To keep up to date with major `sparklyr` announcements, you can follow the RStudio blog (*http://bit.ly/2KQBYVK*).

Community

For general `sparklyr` questions, you can post in the RStudio Community (*http:// bit.ly/2PfNqzN*) tagged as `sparklyr`.

Stack Overflow

For general Spark questions, Stack Overflow (*http://bit.ly/2TEfU4L*) is a great resource; there are also many topics specifically about `sparklyr` (*http://bit.ly/ 307X5cB*).

GitHub

If you believe something needs to be fixed, open a GitHub (*http://bit.ly/ 30b5NGT*) issue or send us a pull request.

Gitter

For urgent issues or to keep in touch, you can chat with us in Gitter (*http://bit.ly/ 33ESccY*).

Recap

In this chapter you learned about the prerequisites required to work with Spark. You saw how to connect to Spark using `spark_connect()`; install a local cluster using `spark_install()`; load a simple dataset, launch the web interface, and display logs using `spark_web(sc)` and `spark_log(sc)`, respectively; and disconnect from RStudio using `spark_disconnect()`. We close by presenting the RStudio extensions that `spar klyr` provides.

At this point, we hope that you feel ready to tackle actual data analysis and modeling problems in Spark and R, which will be introduced over the next two chapters. Chapter 3 will present data analysis as the process of inspecting, cleaning, and transforming data with the goal of discovering useful information. Modeling, the subject of Chapter 4, can be considered part of data analysis; however, it deserves its own chapter to truly describe and take advantage of the modeling functionality available in Spark.

Analysis

First lesson: stick them with the pointy end.

—Jon Snow

Previous chapters focused on introducing Spark with R, getting you up to speed and encouraging you to try basic data analysis workflows. However, they have not properly introduced what data analysis means, especially with Spark. They presented the tools you will need throughout this book—tools that will help you spend more time learning and less time troubleshooting.

This chapter introduces tools and concepts to perform data analysis in Spark from R. Spoiler alert: these are the same tools you use with plain R! This is not a mere coincidence; rather, we want data scientists to live in a world where technology is hidden from them, where you can use the R packages you know and love, and they "just work" in Spark! Now, we are not quite there yet, but we are also not that far. Therefore, in this chapter you learn widely used R packages and practices to perform data analysis—dplyr, ggplot2, formulas, rmarkdown, and so on—which also happen to work in Spark.

Chapter 4 will focus on creating statistical models to predict, estimate, and describe datasets, but first, let's get started with analysis!

Overview

In a data analysis project, the main goal is to understand what the data is trying to "tell us," hoping that it provides an answer to a specific question. Most data analysis projects follow a set of steps, as shown in Figure 3-1.

As the diagram illustrates, we first *import* data into our analysis stem, where we *wrangle* it by trying different data transformations, such as aggregations. We then *visualize* the data to help us perceive relationships and trends. To gain deeper insight, we can fit one or multiple statistical *models* against sample data. This will help us find out whether the patterns hold true when new data is applied to them. Lastly, the results are communicated publicly or privately to colleagues and stakeholders.

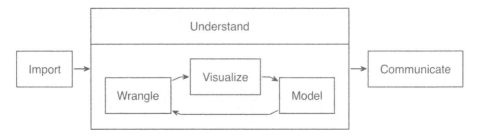

Figure 3-1. The general steps of a data analysis

When working with not-large-scale datasets—as in datasets that fit in memory—we can perform all those steps from R, without using Spark. However, when data does not fit in memory or computation is simply too slow, we can slightly modify this approach by incorporating Spark. But how?

For data analysis, the ideal approach is to let Spark do what it's good at. Spark is a parallel computation engine that works at a large scale and provides a SQL engine and modeling libraries. You can use these to perform most of the same operations R performs. Such operations include data selection, transformation, and modeling. Additionally, Spark includes tools for performing specialized computational work like graph analysis, stream processing, and many others. For now, we will skip those non-rectangular datasets and present them in later chapters.

You can perform data *import*, *wrangling*, and *modeling* within Spark. You can also partly do *visualization* with Spark, which we cover later in this chapter. The idea is to use R to tell Spark what data operations to run, and then only bring the results into R. As illustrated in Figure 3-2, the ideal method *pushes compute* to the Spark cluster and then *collects results* into R.

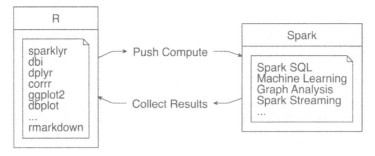

Figure 3-2. Spark computes while R collects results

The `sparklyr` package aids in using the "push compute, collect results" principle. Most of its functions are wrappers on top of Spark API calls. This allows us to take advantage of Spark's analysis components, instead of R's. For example, when you need to fit a linear regression model, instead of using R's familiar `lm()` function, you would use Spark's `ml_linear_regression()` function. This R function then calls Spark to create this model. Figure 3-3 depicts this specific example.

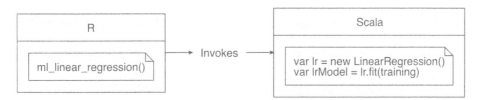

Figure 3-3. R functions call Spark functionality

For more common data manipulation tasks, `sparklyr` provides a backend for `dplyr`. This means you can use `dplyr` verbs with which you're already familiar in R, and then `sparklyr` and `dplyr` will translate those actions into Spark SQL statements, which are generally more compact and easier to read than SQL statements (see Figure 3-4). So, if you are already familiar with R and `dplyr`, there is nothing new to learn. This might feel a bit anticlimactic—indeed, it is—but it's also great since you can focus that energy on learning other skills required to do large-scale computing.

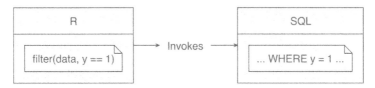

Figure 3-4. dplyr writes SQL in Spark

To practice as you learn, the rest of this chapter's code uses a single exercise that runs in the *local* Spark master. This way, you can replicate the code on your personal computer. Make sure `sparklyr` is already working, which should be the case if you completed Chapter 2.

This chapter will make use of packages that you might not have installed. So, first, make sure the following packages are installed by running these commands:

```
install.packages("ggplot2")
install.packages("corrr")
install.packages("dbplot")
install.packages("rmarkdown")
```

First, load the `sparklyr` and `dplyr` packages and then open a new *local* connection.

```
library(sparklyr)
library(dplyr)

sc <- spark_connect(master = "local", version = "2.3")
```

The environment is ready to be used, so our next task is to import data that we can later analyze.

Import

When using Spark with R, you need to approach importing data differently. Usually, importing means that R will read files and load them into memory; when you are using Spark, the data is imported into Spark, not R. In Figure 3-5, notice how the data source is connected to Spark instead of being connected to R.

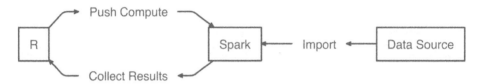

Figure 3-5. Import data to Spark not R

 When you're performing analysis over large-scale datasets, the vast majority of the necessary data will already be available in your Spark cluster (which is usually made available to users via Hive tables or by accessing the file system directly). Chapter 8 will cover this extensively.

Rather than importing all data into Spark, you can request Spark to access the data source without importing it—this is a decision you should make based on speed and performance. Importing all of the data into the Spark session incurs a one-time up-front cost, since Spark needs to wait for the data to be loaded before analyzing it. If the data is not imported, you usually incur a cost with every Spark operation since Spark needs to retrieve a subset from the cluster's storage, which is usually disk drives that happen to be much slower than reading from Spark's memory. More on this topic will be covered in Chapter 9.

Let's prime the session with some data by importing mtcars into Spark using copy_to(); you can also import data from distributed files in many different file formats, which we look at in Chapter 8.

```
cars <- copy_to(sc, mtcars)
```

When using real clusters, you should use copy_to() to transfer only small tables from R; large data transfers should be performed with specialized data transfer tools.

The data is now accessible to Spark and you can now apply transformations with ease; the next section covers how to wrangle data by running transformations inside Spark, using dplyr.

Wrangle

Data wrangling uses transformations to understand the data. It is often referred to as the process of transforming data from one "raw" data form into another format with the intent of making it more appropriate for data analysis.

Malformed or missing values and columns with multiple attributes are common data problems you might need to fix, since they prevent you from understanding your dataset. For example, a "name" field contains the last and first name of a customer. There are two attributes (first and last name) in a single column. To be usable, we need to *transform* the "name" field, by *changing* it into "first_name" and "last_name" fields.

After the data is cleaned, you still need to understand the basics about its content. Other transformations such as aggregations can help with this task. For example, the result of requesting the average balance of all customers will return a single row and column. The value will be the average of all customers. That information will give us context when we see individual, or grouped, customer balances.

The main goal is to write the data transformations using R syntax as much as possible. This saves us from the cognitive cost of having to switch between multiple computer technologies to accomplish a single task. In this case, it is better to take advantage of dplyr instead of writing Spark SQL statements for data exploration.

In the R environment, cars can be treated as if it were a local DataFrame, so you can use dplyr verbs. For instance, we can find out the mean of all columns by using sum marise_all():

```
summarise_all(cars, mean)

# Source: spark<?> [?? x 11]
    mpg   cyl  disp    hp  drat    wt  qsec    vs    am  gear  carb
  <dbl> <dbl> <dbl> <dbl> <dbl> <dbl> <dbl> <dbl> <dbl> <dbl> <dbl>
1  20.1  6.19  231.  147.  3.60  3.22  17.8 0.438 0.406  3.69  2.81
```

While this code is exactly the same as the code you would run when using dplyr without Spark, a lot is happening under the hood. The data is *not* being imported into

R; instead, `dplyr` converts this task into SQL statements that are then sent to Spark. The `show_query()` command makes it possible to peer into the SQL statement that `sparklyr` and `dplyr` created and sent to Spark. We can also use this time to introduce the pipe operator (`%>%`), a custom operator from the `magrittr` package that pipes a computation into the first argument of the next function, making your data analysis much easier to read:

```
summarise_all(cars, mean) %>%
  show_query()

<SQL>
SELECT AVG(`mpg`) AS `mpg`, AVG(`cyl`) AS `cyl`, AVG(`disp`) AS `disp`,
       AVG(`hp`) AS `hp`, AVG(`drat`) AS `drat`, AVG(`wt`) AS `wt`,
       AVG(`qsec`) AS `qsec`, AVG(`vs`) AS `vs`, AVG(`am`) AS `am`,
       AVG(`gear`) AS `gear`, AVG(`carb`) AS `carb`
FROM `mtcars`
```

As is evident, `dplyr` is much more concise than SQL, but rest assured, you will not need to see or understand SQL when using `dplyr`. Your focus can remain on obtaining insights from the data, as opposed to figuring out how to express a given set of transformations in SQL. Here is another example that groups the `cars` dataset by transmission type:

```
cars %>%
  mutate(transmission = ifelse(am == 0, "automatic", "manual")) %>%
  group_by(transmission) %>%
  summarise_all(mean)
```

```
# Source: spark<?> [?? x 12]
  transmission   mpg   cyl  disp    hp  drat    wt  qsec    vs    am  gear  carb
  <chr>        <dbl> <dbl> <dbl> <dbl> <dbl> <dbl> <dbl> <dbl> <dbl> <dbl> <dbl>
1 automatic     17.1  6.95  290.  160.  3.29  3.77  18.2 0.368     0  3.21  2.74
2 manmual       24.4  5.08  144.  127.  4.05  2.41  17.4 0.538     1  4.38  2.92
```

Most of the data transformation operations made available by `dplyr` to work with local DataFrames are also available to use with a Spark connection. This means that you can focus on learning `dplyr` first and then reuse that skill when working with Spark. Chapter 5 from the book *R for Data Science* (*https://r4ds.had.co.nz/*) by Hadley Wickham and Garrett Grolemund (O'Reilly) is a great resource to learn `dplyr` in depth. If proficiency with `dplyr` is not an issue for you, we recommend that you take some time to experiment with different `dplyr` functions against the `cars` table.

Sometimes, we might need to perform an operation not yet available through `dplyr` and `sparklyr`. Instead of downloading the data into R, there is usually a Hive function within Spark to accomplish what we need. The next section covers this scenario.

Built-in Functions

Spark SQL is based on Hive's SQL conventions and functions, and it is possible to call all these functions using `dplyr` as well. This means that we can use any Spark SQL functions to accomplish operations that might not be available via `dplyr`. We can access the functions by calling them as if they were R functions. Instead of failing, `dplyr` passes functions it does not recognize as is to the query engine. This gives us a lot of flexibility on the functions we can use.

For instance, the `percentile()` function returns the exact percentile of a column in a group. The function expects a column name, and either a single percentile value or an array of percentile values. We can use this Spark SQL function from `dplyr`, as follows:

```
summarise(cars, mpg_percentile = percentile(mpg, 0.25))

# Source: spark<?> [?? x 1]
  mpg_percentile
           <dbl>
1           15.4
```

There is no `percentile()` function in R, so `dplyr` passes that portion of the code as-is to the resulting SQL query:

```
summarise(cars, mpg_percentile = percentile(mpg, 0.25)) %>%
  show_query()

<SQL>
SELECT percentile(`mpg`, 0.25) AS `mpg_percentile`
FROM `mtcars_remote`
```

To pass multiple values to `percentile()`, we can call another Hive function called `array()`. In this case, `array()` would work similarly to R's `list()` function. We can pass multiple values separated by commas. The output from Spark is an array variable, which is imported into R as a list variable column:

```
summarise(cars, mpg_percentile = percentile(mpg, array(0.25, 0.5, 0.75)))

# Source: spark<?> [?? x 1]
  mpg_percentile
  <list>
1 <list [3]>
```

You can use the `explode()` function to separate Spark's array value results into their own record. To do this, use `explode()` within a `mutate()` command, and pass the variable containing the results of the percentile operation:

```
summarise(cars, mpg_percentile = percentile(mpg, array(0.25, 0.5, 0.75))) %>%
  mutate(mpg_percentile = explode(mpg_percentile))

# Source: spark<?> [?? x 1]
  mpg_percentile
           <dbl>
```

```
1           15.4
2           19.2
3           22.8
```

We have included a comprehensive list of all the Hive functions in the section "Hive Functions" on page 255. Glance over them to get a sense of the wide range of operations that you can accomplish with them.

Correlations

A very common exploration technique is to calculate and visualize correlations, which we often calculate to find out what kind of statistical relationship exists between paired sets of variables. Spark provides functions to calculate correlations across the entire dataset and returns the results to R as a DataFrame object:

```
ml_corr(cars)

# A tibble: 11 x 11
      mpg    cyl   disp     hp   drat     wt   qsec
    <dbl>  <dbl>  <dbl>  <dbl>  <dbl>  <dbl>  <dbl>
 1  1     -0.852 -0.848 -0.776  0.681 -0.868  0.419
 2 -0.852  1      0.902  0.832 -0.700  0.782 -0.591
 3 -0.848  0.902  1      0.791 -0.710  0.888 -0.434
 4 -0.776  0.832  0.791  1     -0.449  0.659 -0.708
 5  0.681 -0.700 -0.710 -0.449  1     -0.712  0.0912
 6 -0.868  0.782  0.888  0.659 -0.712  1     -0.175
 7  0.419 -0.591 -0.434 -0.708  0.0912 -0.175  1
 8  0.664 -0.811 -0.710 -0.723  0.440 -0.555  0.745
 9  0.600 -0.523 -0.591 -0.243  0.713 -0.692 -0.230
10  0.480 -0.493 -0.556 -0.126  0.700 -0.583 -0.213
11 -0.551  0.527  0.395  0.750 -0.0908 0.428 -0.656
# ... with 4 more variables: vs <dbl>, am <dbl>,
#   gear <dbl>, carb <dbl>
```

The corrr R package specializes in correlations. It contains friendly functions to prepare and visualize the results. Included inside the package is a backend for Spark, so when a Spark object is used in corrr, the actual computation also happens in Spark. In the background, the correlate() function runs sparklyr::ml_corr(), so there is no need to collect any data into R prior to running the command:

```
library(corrr)
correlate(cars, use = "pairwise.complete.obs", method = "pearson")

# A tibble: 11 x 12
    rowname    mpg    cyl   disp     hp   drat     wt
    <chr>    <dbl>  <dbl>  <dbl>  <dbl>  <dbl>  <dbl>
 1 mpg      NA     -0.852 -0.848 -0.776  0.681 -0.868
 2 cyl      -0.852 NA      0.902  0.832 -0.700  0.782
 3 disp     -0.848  0.902 NA      0.791 -0.710  0.888
 4 hp       -0.776  0.832  0.791 NA     -0.449  0.659
 5 drat      0.681 -0.700 -0.710 -0.449 NA     -0.712
 6 wt       -0.868  0.782  0.888  0.659 -0.712 NA
```

```
 7 qsec     0.419  -0.591  -0.434  -0.708   0.0912  -0.175
 8 vs       0.664  -0.811  -0.710  -0.723   0.440   -0.555
 9 am       0.600  -0.523  -0.591  -0.243   0.713   -0.692
10 gear     0.480  -0.493  -0.556  -0.126   0.700   -0.583
11 carb    -0.551   0.527   0.395   0.750  -0.0908   0.428
# ... with 5 more variables: qsec <dbl>, vs <dbl>,
#   am <dbl>, gear <dbl>, carb <dbl>
```

We can pipe the results to other corrr functions. For example, the shave() function turns all of the duplicated results into NAs. Again, while this feels like standard R code using existing R packages, Spark is being used under the hood to perform the correlation.

Additionally, as shown in Figure 3-6, the results can be easily visualized using the rplot() function, as shown here:

```
correlate(cars, use = "pairwise.complete.obs", method = "pearson") %>%
  shave() %>%
  rplot()
```

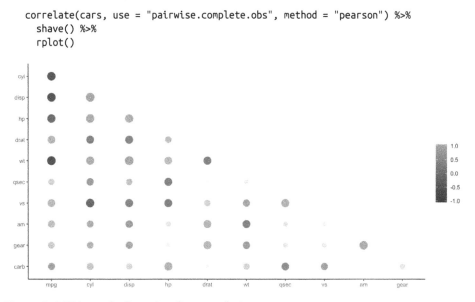

Figure 3-6. Using rplot() to visualize correlations

It is much easier to see which relationships are positive or negative: positive relationships are in gray, and negative relationships are black. The size of the circle indicates how significant their relationship is. The power of visualizing data is in how much easier it makes it for us to understand results. The next section expands on this step of the process.

Visualize

Visualizations are a vital tool to help us find patterns in the data. It is easier for us to identify outliers in a dataset of 1,000 observations when plotted in a graph, as opposed to reading them from a list.

R is great at data visualizations. Its capabilities for creating plots are extended by the many R packages that focus on this analysis step. Unfortunately, the vast majority of R functions that create plots depend on the data already being in local memory within R, so they fail when using a remote table within Spark.

It is possible to create visualizations in R from data sources that exist in Spark. To understand how to do this, let's first break down how computer programs build plots. To begin, a program takes the raw data and performs some sort of transformation. The transformed data is then mapped to a set of coordinates. Finally, the mapped values are drawn in a plot. Figure 3-7 summarizes each of the steps.

Figure 3-7. Stages of an R plot

In essence, the approach for visualizing is the same as in wrangling: push the computation to Spark, and then collect the results in R for plotting. As illustrated in Figure 3-8, the heavy lifting of preparing the data, such as aggregating the data by groups or bins, can be done within Spark, and then the much smaller dataset can be collected into R. Inside R, the plot becomes a more basic operation. For example, for a histogram, the bins are calculated in Spark, and then plotted in R using a simple column plot, as opposed to a histogram plot, because there is no need for R to recalculate the bins.

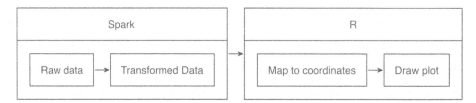

Figure 3-8. Plotting with Spark and R

Let's apply this conceptual model when using ggplot2.

Using ggplot2

To create a bar plot using ggplot2, we simply call a function:

```
library(ggplot2)
ggplot(aes(as.factor(cyl), mpg), data = mtcars) + geom_col()
```

In this case, the `mtcars` raw data was *automatically* transformed into three discrete aggregated numbers. Next, each result was mapped into an x and y plane. Then the plot was drawn. As R users, all of the stages of building the plot are conveniently abstracted for us.

In Spark, there are a couple of key steps when codifying the "push compute, collect results" approach. First, ensure that the transformation operations happen within Spark. In the example that follows, `group_by()` and `summarise()` will run inside Spark. The second is to bring the results back into R after the data has been transformed. Be sure to transform and then collect, in that order; if `collect()` is run first, R will try to ingest the entire dataset from Spark. Depending on the size of the data, collecting all of the data will slow down or can even bring down your system.

```
car_group <- cars %>%
  group_by(cyl) %>%
  summarise(mpg = sum(mpg, na.rm = TRUE)) %>%
  collect() %>%
  print()

# A tibble: 3 x 2
    cyl   mpg
  <dbl> <dbl>
1     6  138.
2     4  293.
3     8  211.
```

In this example, now that the data has been preaggregated and collected into R, only three records are passed to the plotting function:

```
ggplot(aes(as.factor(cyl), mpg), data = car_group) +
  geom_col(fill = "#999999") + coord_flip()
```

Figure 3-9 shows the resulting plot.

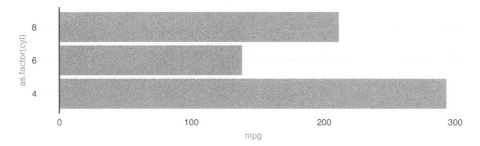

Figure 3-9. Plot with aggregation in Spark

Any other `ggplot2` visualization can be made to work using this approach; however, this is beyond the scope of the book. Instead, we recommend that you read *R Graphics Cookbook* (*https://oreil.ly/bIF4*), by Winston Chang (O'Reilly) to learn additional visualization techniques applicable to Spark. Now, to ease this transformation step before visualizing, the `dbplot` package provides a few ready-to-use visualizations that automate aggregation in Spark.

Using dbplot

The `dbplot` package provides helper functions for plotting with remote data. The R code `dbplot` that's used to transform the data is written so that it can be translated into Spark. It then uses those results to create a graph using the `ggplot2` package where data transformation and plotting are both triggered by a single function.

The `dbplot_histogram()` function makes Spark calculate the bins and the count per bin and outputs a `ggplot` object, which we can further refine by adding more steps to the plot object. `dbplot_histogram()` also accepts a `binwidth` argument to control the range used to compute the bins:

```
library(dbplot)

cars %>%
dbplot_histogram(mpg, binwidth = 3) +
labs(title = "MPG Distribution",
     subtitle = "Histogram over miles per gallon")
```

Figure 3-10 presents the resulting plot.

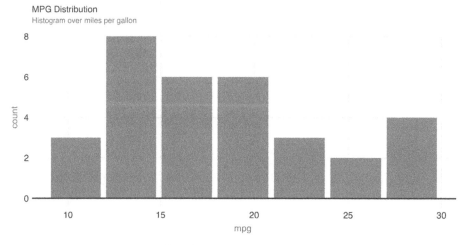

Figure 3-10. Histogram created by dbplot

Histograms provide a great way to analyze a single variable. To analyze two variables, a scatter or raster plot is commonly used.

Scatter plots are used to compare the relationship between two continuous variables. For example, a scatter plot will display the relationship between the weight of a car and its gas consumption. The plot in Figure 3-11 shows that the higher the weight, the higher the gas consumption because the dots clump together into almost a line that goes from the upper left toward the lower right. Here's the code to generate the plot:

```
ggplot(aes(mpg, wt), data = mtcars) +
  geom_point()
```

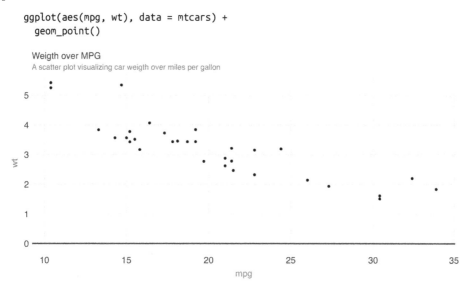

Figure 3-11. Scatter plot example in Spark

However, for scatter plots, no amount of "pushing the computation" to Spark will help with this problem because the data must be plotted in individual dots.

The best alternative is to find a plot type that represents the x/y relationship and concentration in a way that it is easy to perceive and to "physically" plot. The *raster* plot might be the best answer. A raster plot returns a grid of x/y positions and the results of a given aggregation, usually represented by the color of the square.

You can use `dbplot_raster()` to create a scatter-like plot in Spark, while only retrieving (collecting) a small subset of the remote dataset:

```
dbplot_raster(cars, mpg, wt, resolution = 16)
```

As shown in Figure 3-12, the resulting plot returns a grid no bigger than 5 x 5. This limits the number of records that need to be collected into R to 25.

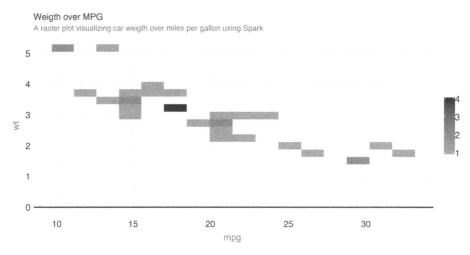

Figure 3-12. A raster plot using Spark

 You can also use `dbplot` to retrieve the raw data and visualize by other means; to retrieve the aggregates, but not the plots, use `db_compute_bins()`, `db_compute_count()`, `db_compute_raster()`, and `db_compute_boxplot()`.

While visualizations are indispensable, you can complement data analysis using statistical models to gain even deeper insights into our data. The next section describes how we can prepare data for modeling with Spark.

Model

The next two chapters focus entirely on modeling, so rather than introducing modeling in too much detail in this chapter, we want to cover how to interact with models while doing data analysis.

First, an analysis project goes through many transformations and models to find the answer. That's why the first data analysis diagram we introduced in Figure 3-2 illustrates a cycle of: visualizing, wrangling, and modeling—we know you don't end with modeling, not in R, nor when using Spark.

Therefore, the ideal data analysis language enables you to quickly adjust over each wrangle-visualize-model iteration. Fortunately, this is the case when using Spark and R.

To illustrate how easy it is to iterate over wrangling and modeling in Spark, consider the following example. We will start by performing a linear regression against all features and predict miles per gallon:

```
cars %>%
  ml_linear_regression(mpg ~ .) %>%
  summary()

Deviance Residuals:
    Min      1Q   Median      3Q      Max
-3.4506  -1.6044  -0.1196   1.2193   4.6271

Coefficients:
(Intercept)         cyl         disp          hp         drat           wt
12.30337416  -0.11144048   0.01333524  -0.02148212   0.78711097  -3.71530393
       qsec          vs           am         gear         carb
 0.82104075   0.31776281   2.52022689   0.65541302  -0.19941925

R-Squared: 0.869
Root Mean Squared Error: 2.147
```

At this point, it is very easy to experiment with different features, we can simply change the R formula from `mpg ~ .` to, say, `mpg ~ hp + cyl` to only use horsepower and cylinders as features:

```
cars %>%
  ml_linear_regression(mpg ~ hp + cyl) %>%
  summary()

Deviance Residuals:
    Min      1Q   Median      3Q      Max
-4.4948  -2.4901  -0.1828   1.9777   7.2934

Coefficients:
(Intercept)          hp          cyl
 36.9083305  -0.0191217   -2.2646936

R-Squared: 0.7407
Root Mean Squared Error: 3.021
```

Additionally, it is also very easy to iterate with other kinds of models. The following one replaces the linear model with a generalized linear model:

```
cars %>%
  ml_generalized_linear_regression(mpg ~ hp + cyl) %>%
  summary()

Deviance Residuals:
    Min      1Q   Median      3Q      Max
-4.4948  -2.4901  -0.1828   1.9777   7.2934

Coefficients:
(Intercept)          hp          cyl
 36.9083305  -0.0191217   -2.2646936

(Dispersion parameter for gaussian family taken to be 10.06809)

    Null  deviance: 1126.05 on 31 degress of freedom
```

```
Residual deviance: 291.975 on 29 degrees of freedom
AIC: 169.56
```

Usually, before fitting a model you would need to use multiple dplyr transformations to get it ready to be consumed by a model. To make sure the model can be fitted as efficiently as possible, you should cache your dataset before fitting it, as described next.

Caching

The examples in this chapter are built using a very small dataset. In real-life scenarios, large amounts of data are used for models. If the data needs to be transformed first, the volume of the data could exact a heavy toll on the Spark session. Before fitting the models, it is a good idea to save the results of all the transformations in a new table loaded in Spark memory.

The compute() command can take the end of a dplyr command and save the results to Spark memory:

```
cached_cars <- cars %>%
  mutate(cyl = paste0("cyl_", cyl)) %>%
  compute("cached_cars")

cached_cars %>%
  ml_linear_regression(mpg ~ .) %>%
  summary()

Deviance Residuals:
    Min       1Q    Median       3Q       Max
-3.47339 -1.37936 -0.06554  1.05105  4.39057

Coefficients:
(Intercept) cyl_cyl_8.0 cyl_cyl_4.0        disp          hp        drat
16.15953652  3.29774653  1.66030673  0.01391241 -0.04612835  0.02635025
         wt        qsec          vs          am        gear        carb
-3.80624757  0.64695710  1.74738689  2.61726546 0.76402917  0.50935118

R-Squared: 0.8816
Root Mean Squared Error: 2.041
```

As more insights are gained from the data, more questions might be raised. That is why we expect to iterate through the data wrangle, visualize, and model cycle multiple times. Each iteration should provide incremental insights into what the data is "telling us." There will be a point when we reach a satisfactory level of understanding. It is at this point that we will be ready to share the results of the analysis. This is the topic of the next section.

Communicate

It is important to clearly communicate the analysis results—as important as the analysis work itself! The public, colleagues, or stakeholders need to understand what you found out and how.

To communicate effectively, we need to use artifacts such as reports and presentations; these are common output formats that we can create in R, using *R Markdown*.

R Markdown documents allow you to weave narrative text and code together. The variety of output formats provides a very compelling reason to learn and use R Markdown. There are many available output formats like HTML, PDF, PowerPoint, Word, web slides, websites, books, and so on.

Most of these outputs are available in the core R packages of R Markdown: `knitr` and `rmarkdown`. You can extend R Markdown with other R packages. For example, this book was written using R Markdown thanks to an extension provided by the `book down` package. The best resource to delve deeper into R Markdown is the official book.[1]

In R Markdown, one singular artifact could potentially be rendered in different formats. For example, you could render the same report in HTML or as a PDF file by changing a setting within the report itself. Conversely, multiple types of artifacts could be rendered as the same output. For example, a presentation deck and a report could be rendered in HTML.

Creating a new R Markdown report that uses Spark as a computer engine is easy. At the top, R Markdown expects a YAML header. The first and last lines are three consecutive dashes (`---`). The content in between the dashes varies depending on the type of document. The only required field in the YAML header is the `output` value. R Markdown needs to know what kind of output it needs to render your report into. This YAML header is called *frontmatter*. Following the frontmatter are sections of code, called *code chunks*. These code chunks can be interlaced with the narratives. There is nothing particularly interesting to note when using Spark with R Markdown; it is just business as usual.

Since an R Markdown document is self-contained and meant to be reproducible, before rendering documents, we should first disconnect from Spark to free resources:

```
spark_disconnect(sc)
```

The following example shows how easy it is to create a fully reproducible report that uses Spark to process large-scale datasets. The narrative, code, and, most important,

---

1 Xie Allaire G (2018). *R Markdown: The Definite Guide*, 1st edition. CRC Press.

the output of the code is recorded in the resulting HTML file. You can copy and paste the following code in a file. Save the file with a *.Rmd* extension, and choose whatever name you would like:

```
---
title: "mtcars analysis"
output:
  html_document:
    fig_width: 6
    fig_height: 3
---
```{r, setup, include = FALSE}
library(sparklyr)
library(dplyr)

sc <- spark_connect(master = "local", version = "2.3")
cars <- copy_to(sc, mtcars)
```


Visualize
Aggregate data in Spark, visualize in R.
```{r  fig.align='center', warning=FALSE}
library(ggplot2)
cars %>%
  group_by(cyl) %>% summarise(mpg = mean(mpg)) %>%
  ggplot(aes(cyl, mpg)) + geom_bar(stat="identity")
```


Model
The selected model was a simple linear regression that
uses the weight as the predictor of MPG

```{r}
cars %>%
  ml_linear_regression(wt ~ mpg) %>%
  summary()
```
```{r, include = FALSE}
spark_disconnect(sc)
```
```

To knit this report, save the file with a *.Rmd* extension such as *report.Rmd*, and run render() from R. The output should look like that shown in Figure 3-13.

```
rmarkdown::render("report.Rmd")
```

mtcars analysis

Visualize

Aggregate data in Spark, visualize in R.

```
library(ggplot2)
cars %>%
  group_by(cyl) %>% summarise(mpg = mean(mpg)) %>%
  ggplot(aes(cyl, mpg)) + geom_bar(stat="identity")
```

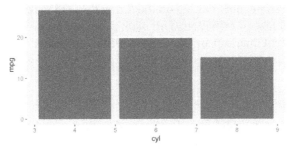

Model

The selected model was a simple linear regression that uses the weight as the predictor of MPG

```
cars %>%
  ml_linear_regression(wt ~ mpg) %>%
  summary()
```

```
## Deviance Residuals:
##     Min      1Q  Median      3Q     Max
## -0.6516 -0.3490 -0.1381  0.3190  1.3684
##
## Coefficients:
## (Intercept)         mpg
##    6.047255   -0.140862
##
## R-Squared: 0.7528
## Root Mean Squared Error: 0.4788
```

Figure 3-13. R Markdown HTML output

You can now easily share this report, and viewers of won't need Spark or R to read
and consume its contents; it's just a self-contained HTML file, trivial to open in any
browser.

It is also common to distill insights of a report into many other output formats.
Switching is quite easy: in the top frontmatter, change the output option to power
point_presentation, pdf_document, word_document, or the like. Or you can even
produce multiple output formats from the same report:

```
---
title: "mtcars analysis"
output:
  word_document: default
  pdf_document: default
```

```
powerpoint_presentation: default
---
```

The result will be a PowerPoint presentation, a Word document, and a PDF. All of the same information that was displayed in the original HTML report is computed in Spark and rendered in R. You'll likely need to edit the PowerPoint template or the output of the code chunks.

This minimal example shows how easy it is to go from one format to another. Of course, it will take some more editing on the R user's side to make sure the slides contain only the pertinent information. The main point is that it does not require that you learn a different markup or code conventions to switch from one artifact to another.

Recap

This chapter presented a solid introduction to data analysis with R and Spark. Many of the techniques presented looked quite similar to using just R and no Spark, which, while anticlimactic, is the right design to help users already familiar with R to easily transition to Spark. For users unfamiliar with R, this chapter also served as a very brief introduction to some of the most popular (and useful!) packages available in R.

It should now be quite obvious that, together, R and Spark are a powerful combination—a large-scale computing platform, along with an incredibly robust ecosystem of R packages, makes for an ideal analysis platform.

While doing analysis in Spark with R, remember to push computation to Spark and focus on collecting results in R. This paradigm should set up a successful approach to data manipulation, visualization and communication through sharing your results in a variety of outputs.

Chapter 4 will dive deeper into how to build statistical models in Spark using a much more interesting dataset (what's more interesting than dating datasets?). You will also learn many more techniques that we did not even mention in the brief modeling section from this chapter.

Modeling

I've trusted in your visions, in your prophecies, for
years.

—*Stannis Baratheon*

In Chapter 3 you learned how to scale up data analysis to large datasets using Spark. In this chapter, we detail the steps required to build prediction models in Spark. We explore `MLlib`, the component of Spark that allows you to write high-level code to perform predictive modeling on distributed data, and use data wrangling in the context of feature engineering and exploratory data analysis.

We will start this chapter by introducing modeling in the context of Spark and the dataset you will use throughout the chapter. We then demonstrate a supervised learning workflow that includes exploratory data analysis, feature engineering, and model building. Then we move on to an unsupervised topic modeling example using unstructured text data. Keep in mind that our goal is to show various techniques of executing data science tasks on large data rather than conducting a rigorous and coherent analysis. There are also many other models available in Spark that won't be covered in this chapter, but by the end of the chapter, you will have the right tools to experiment with additional ones on your own.

While predicting datasets manually is often a reasonable approach (by "manually," we mean someone imports a dataset into Spark and uses the fitted model to enrich or predict values), it does beg the question, could we automate this process into systems that anyone can use? For instance, how can we build a system that automatically identifies an email as spam without having to manually analyze each email account? Chapter 5 presents the tools to automate data analysis and modeling with pipelines, but to get there, we need to first understand how to train models "by hand."

Overview

The R interface to Spark provides modeling algorithms that should be familiar to R users. For instance, we've already used `ml_linear_regression(cars, mpg ~ .)`, but we could just as easily run `ml_logistic_regression(cars, am ~ .)`.

Take a moment to look at the long list of `MLlib` functions included in the appendix of this book; a quick glance at this list shows that Spark supports Decision Trees, Gradient-Boosted Trees, Accelerated Failure Time Survival Regression, Isotonic Regression, *K*-Means Clustering, Gaussian Mixture Clustering, and more.

As you can see, Spark provides a wide range of algorithms and feature transformers, and here we touch on a representative portion of the functionality. A complete treatment of predictive modeling concepts is beyond the scope of this book, so we recommend complementing this discussion with *R for Data Science* (*https://r4ds.had.co.nz/*) by Hadley Wickham and Garrett Grolemund G (O'Reilly) and *Feature Engineering and Selection: A Practical Approach for Predictive Models*,[1] from which we adopted (sometimes verbatim) some of the examples and visualizations in this chapter.

This chapter focuses on predictive modeling, since Spark aims to enable machine learning as opposed to statistical inference. Machine learning is often more concerned about forecasting the future rather than inferring the process by which our data is generated,[2] which is then used to create automated systems. Machine learning can be categorized into *supervised learning* (predictive modeling) and *unsupervised learning*. In supervised learning, we try to learn a function that will map from X to Y, from a dataset of (x, y) examples. In unsupervised learning, we just have X and not the Y labels, so instead we try to learn something about the structure of X. Some practical use cases for supervised learning include forecasting tomorrow's weather, determining whether a credit card transaction is fraudulent, and coming up with a quote for your car insurance policy. With unsupervised learning, examples include automated grouping of photos of individuals, segmenting customers based on their purchase history, and clustering of documents.

The ML interface in `sparklyr` has been designed to minimize the cognitive effort for moving from a local, in-memory, native-R workflow to the cluster, and back. While the Spark ecosystem is very rich, there is still a tremendous number of packages from CRAN, with some implementing functionality that you might require for a project. Also, you might want to leverage your skills and experience working in R to maintain productivity. What we learned in Chapter 3 also applies here—it is important to keep

---

1 Kuhn M, Johnson K (2019). *Feature Engineering and Selection: A Practical Approach for Predictive Models.* (CRC PRess.)

2 We acknowledge that the terms here might mean different things to different people and that there is a continuum between the two approaches; however, they are defined.

---

track of where you are performing computations and move between the cluster and your R session as appropriate.

The examples in this chapter utilize the OkCupid dataset (*https://oreil.ly/Uv9r_*).[3] The dataset consists of user profile data from an online dating site and contains a diverse set of features, including biographical characteristics such as gender and profession, as well as free text fields related to personal interests. There are about 60,000 profiles in the dataset, which fits comfortably into memory on a modern laptop and wouldn't be considered "big data," so you can easily follow along running Spark in local mode.

You can download this dataset as follows:

```
download.file(
  "https://github.com/r-spark/okcupid/raw/master/profiles.csv.zip",
  "okcupid.zip")

unzip("okcupid.zip", exdir = "data")
unlink("okcupid.zip")
```

We don't recommend sampling this dataset since the model won't be nearly as rich; however, if you have limited hardware resources, you are welcome to sample it as follows:

```
profiles <- read.csv("data/profiles.csv")
write.csv(dplyr::sample_n(profiles, 10^3),
          "data/profiles.csv", row.names = FALSE)
```

 The examples in this chapter utilize small datasets so that you can easily follow along in local mode. In practice, if your dataset fits comfortably in memory on your local machine, you might be better off using an efficient, nondistributed implementation of the modeling algorithm. For example, you might want to use the ranger package instead of ml_random_forest_classifier().

In addition, to follow along, you will need to install a few additional packages:

```
install.packages("ggmosaic")
install.packages("forcats")
install.packages("FactoMineR")
```

To motivate the examples, we consider the following problem:

> Predict whether someone is actively working—that is, not retired, a student, or unemployed.

Next up, we explore this dataset.

---

3 Kim AY, Escobedo-Land A (2015). "OKCupid data for introductory statistics and data science courses." *Journal of Statistics Education*, 23(2).

Exploratory Data Analysis

Exploratory data analysis (EDA), in the context of predictive modeling, is the exercise of looking at excerpts and summaries of the data. The specific goals of the EDA stage are informed by the business problem, but here are some common objectives:

- Check for data quality; confirm meaning and prevalence of missing values and reconcile statistics against existing controls.

- Understand univariate relationships between variables.

- Perform an initial assessment on what variables to include and what transformations need to be done on them.

To begin, we connect to Spark, load libraries, and read in the data:

```
library(sparklyr)
library(ggplot2)
library(dbplot)
library(dplyr)

sc <- spark_connect(master = "local", version = "2.3")

okc <- spark_read_csv(
  sc,
  "data/profiles.csv",
  escape = "\"",
  memory = FALSE,
  options = list(multiline = TRUE)
) %>%
  mutate(
    height = as.numeric(height),
    income = ifelse(income == "-1", NA, as.numeric(income))
  ) %>%
  mutate(sex = ifelse(is.na(sex), "missing", sex)) %>%
  mutate(drinks = ifelse(is.na(drinks), "missing", drinks)) %>%
  mutate(drugs = ifelse(is.na(drugs), "missing", drugs)) %>%
  mutate(job = ifelse(is.na(job), "missing", job))
```

We specify `escape = "\""` and `options = list(multiline = TRUE)` here to accommodate embedded quote characters and newlines in the essay fields. We also convert the `height` and `income` columns to numeric types and recode missing values in the string columns. Note that it might very well take a few tries of specifying different parameters to get the initial data ingest correct, and sometimes you might need to revisit this step after you learn more about the data during modeling.

We can now take a quick look at our data by using `glimpse()`:

```
glimpse(okc)
```

```
Observations: ??
Variables: 31
Database: spark_connection
$ age         <int> 22, 35, 38, 23, 29, 29, 32, 31, 24, 37, 35…
$ body_type   <chr> "a little extra", "average", "thin", "thin…
$ diet        <chr> "strictly anything", "mostly other", "anyt…
$ drinks      <chr> "socially", "often", "socially", "socially…
$ drugs       <chr> "never", "sometimes", "missing", "missing"…
$ education   <chr> "working on college/university", "working …
$ essay0      <chr> "about me:<br />\n<br />\ni would love to …
$ essay1      <chr> "currently working as an international age…
$ essay2      <chr> "making people laugh.<br />\nranting about…
$ essay3      <chr> "the way i look. i am a six foot half asia…
$ essay4      <chr> "books:<br />\nabsurdistan, the republic, …
$ essay5      <chr> "food.<br />\nwater.<br />\ncell phone.<br…
$ essay6      <chr> "duality and humorous things", "missing", …
$ essay7      <chr> "trying to find someone to hang out with. …
$ essay8      <chr> "i am new to california and looking for so…
$ essay9      <chr> "you want to be swept off your feet!<br />…
$ ethnicity   <chr> "asian, white", "white", "missing", "white…
$ height      <dbl> 75, 70, 68, 71, 66, 67, 65, 65, 67, 65, 70…
$ income      <dbl> NaN, 80000, NaN, 20000, NaN, NaN, NaN, NaN…
$ job         <chr> "transportation", "hospitality / travel", …
$ last_online <chr> "2012-06-28-20-30", "2012-06-29-21-41", "2…
$ location    <chr> "south san francisco, california", "oaklan…
$ offspring   <chr> "doesn&rsquo;t have kids, but might want t…
$ orientation <chr> "straight", "straight", "straight", "strai…
$ pets        <chr> "likes dogs and likes cats", "likes dogs a…
$ religion    <chr> "agnosticism and very serious about it", "…
$ sex         <chr> "m", "m", "m", "m", "m", "m", "f", "f", "f…
$ sign        <chr> "gemini", "cancer", "pisces but it doesn&r…
$ smokes      <chr> "sometimes", "no", "no", "no", "no", "no",…
$ speaks      <chr> "english", "english (fluently), spanish (p…
$ status      <chr> "single", "single", "available", "single",…
```

Now, we add our response variable as a column in the dataset and look at its distribution:

```
okc <- okc %>%
  mutate(
    not_working = ifelse(job %in% c("student", "unemployed", "retired"), 1 , 0)
  )

okc %>%
  group_by(not_working) %>%
  tally()

# Source: spark<?> [?? x 2]
  not_working     n
        <dbl> <dbl>
1           0 54541
2           1  5405
```

Before we proceed further, let's perform an initial split of our data into a training set and a testing set and put away the latter. In practice, this is a crucial step because we would like to have a holdout set that we set aside at the end of the modeling process to evaluate model performance. If we were to include the entire dataset during EDA, information from the testing set could "leak" into the visualizations and summary statistics and bias our model-building process even though the data is not used directly in a learning algorithm. This would undermine the credibility of our performance metrics. We can easily split the data by using the sdf_random_split() function:

```
data_splits <- sdf_random_split(okc, training = 0.8, testing = 0.2, seed = 42)
okc_train <- data_splits$training
okc_test <- data_splits$testing
```

We can quickly look at the distribution of our response variable:

```
okc_train %>%
  group_by(not_working) %>%
  tally() %>%
  mutate(frac = n / sum(n))

# Source: spark<?> [?? x 3]
  not_working     n   frac
        <dbl> <dbl>  <dbl>
1           0 43785 0.910
2           1  4317 0.0897
```

Using the sdf_describe() function, we can obtain numerical summaries of specific columns:

```
sdf_describe(okc_train, cols = c("age", "income"))

# Source: spark<?> [?? x 3]
  summary age                 income
  <chr>   <chr>               <chr>
1 count   48102               9193
2 mean    32.336534863415245  104968.99815076689
3 stddev  9.43908920033797    202235.2291773537
4 min     18                  20000.0
5 max     110                 1000000.0
```

Like we saw in Chapter 3, we can also utilize the dbplot package to plot distributions of these variables. In Figure 4-1 we show a histogram of the distribution of the age variable, which is the result of the following code:

```
dbplot_histogram(okc_train, age)
```

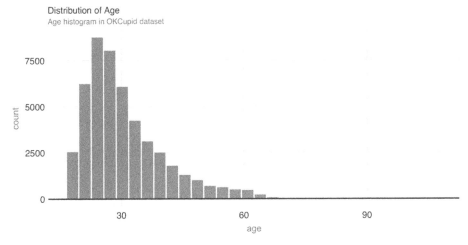

Figure 4-1. Distribution of age

A common EDA exercise is to look at the relationships between the response and the individual predictors. Often, you might have prior business knowledge of what these relationships should be, so this can serve as a data quality check. Also, unexpected trends can inform variable interactions that you might want to include in the model. As an example, we can explore the `religion` variable:

```
prop_data <- okc_train %>%
  mutate(religion = regexp_extract(religion, "^\\\\w+", 0)) %>%
  group_by(religion, not_working) %>%
  tally() %>%
  group_by(religion) %>%
  summarise(
    count = sum(n),
    prop = sum(not_working * n) / sum(n)
  ) %>%
  mutate(se = sqrt(prop * (1 - prop) / count)) %>%
  collect()

prop_data

# A tibble: 10 x 4
   religion     count   prop     se
   <chr>        <dbl>  <dbl>  <dbl>
 1 judaism       2520 0.0794 0.00539
 2 atheism       5624 0.118  0.00436
 3 christianity  4671 0.120  0.00480
 4 hinduism       358 0.101  0.0159
 5 islam          115 0.191  0.0367
 6 agnosticism   7078 0.0958 0.00346
 7 other         6240 0.0841 0.00346
 8 missing      16152 0.0719 0.002
```

```
 9 buddhism      1575 0.0851 0.007
10 catholicism   3769 0.0886 0.00458
```

Note that `prop_data` is a small DataFrame that has been collected into memory in our R session, we can take advantage of `ggplot2` to create an informative visualization (see Figure 4-2):

```
prop_data %>%
  ggplot(aes(x = religion, y = prop)) + geom_point(size = 2) +
  geom_errorbar(aes(ymin = prop - 1.96 * se, ymax = prop + 1.96 * se),
              width = .1) +
  geom_hline(yintercept = sum(prop_data$prop * prop_data$count) /
                          sum(prop_data$count))
```

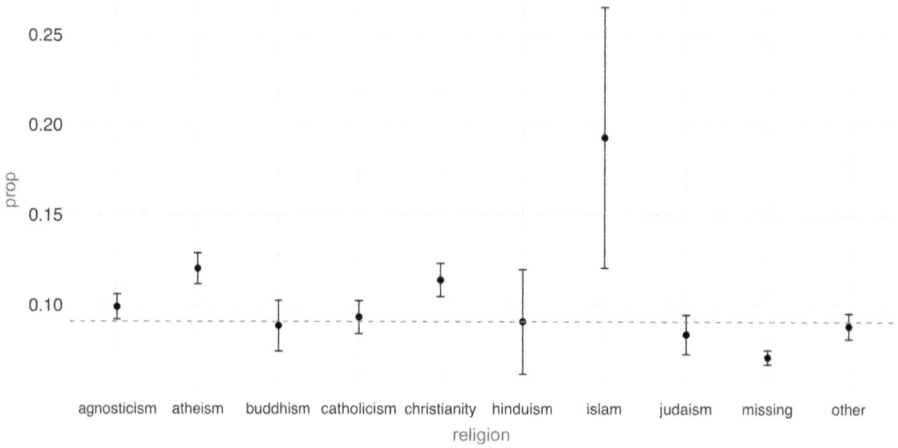

Figure 4-2. Proportion of individuals not currently employed, by religion

Next, we take a look at the relationship between a couple of predictors: alcohol use and drug use. We would expect there to be some correlation between them. You can compute a contingency table via `sdf_crosstab()`:

```
contingency_tbl <- okc_train %>%
  sdf_crosstab("drinks", "drugs") %>%
  collect()

contingency_tbl

# A tibble: 7 x 5
  drinks_drugs missing never often sometimes
  <chr>          <dbl> <dbl> <dbl>     <dbl>
1 very often        54   144    44       137
2 socially        8221 21066   126      4106
3 not at all       146  2371    15       109
4 desperately       72    89    23        74
5 often           1049  1718    69      1271
```

```
6 missing          1121  1226    10        59
7 rarely            613  3689    35       445
```

We can visualize this contingency table using a mosaic plot (see Figure 4-3):

```
library(ggmosaic)
library(forcats)
library(tidyr)

contingency_tbl %>%
  rename(drinks = drinks_drugs) %>%
  gather("drugs", "count", missing:sometimes) %>%
  mutate(
    drinks = as_factor(drinks) %>%
      fct_relevel("missing", "not at all", "rarely", "socially",
                  "very often", "desperately"),
    drugs = as_factor(drugs) %>%
      fct_relevel("missing", "never", "sometimes", "often")
  ) %>%
  ggplot() +
  geom_mosaic(aes(x = product(drinks, drugs), fill = drinks,
                  weight = count))
```

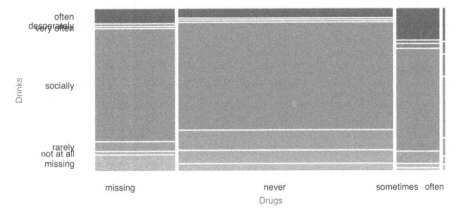

Figure 4-3. Mosaic plot of drug and alcohol use

To further explore the relationship between these two variables, we can perform correspondence analysis[4] using the FactoMineR package. This technique enables us to summarize the relationship between the high-dimensional factor levels by mapping each level to a point on the plane. We first obtain the mapping using Facto MineR::CA() as follows:

---

4 Greenacre M (2017). *Correspondence analysis in practice.* Chapman and Hall/CRC.

```
dd_obj <- contingency_tbl %>%
  tibble::column_to_rownames(var = "drinks_drugs") %>%
  FactoMineR::CA(graph = FALSE)
```

We can then plot the results using `ggplot`, which you can see in Figure 4-4:

```
dd_drugs <-
  dd_obj$row$coord %>%
  as.data.frame() %>%
  mutate(
    label = gsub("_", " ", rownames(dd_obj$row$coord)),
    Variable = "Drugs"
  )

dd_drinks <-
  dd_obj$col$coord %>%
  as.data.frame() %>%
  mutate(
    label = gsub("_", " ", rownames(dd_obj$col$coord)),
    Variable = "Alcohol"
  )

ca_coord <- rbind(dd_drugs, dd_drinks)

ggplot(ca_coord, aes(x = `Dim 1`, y = `Dim 2`,
                     col = Variable)) +
  geom_vline(xintercept = 0) +
  geom_hline(yintercept = 0) +
  geom_text(aes(label = label)) +
  coord_equal()
```

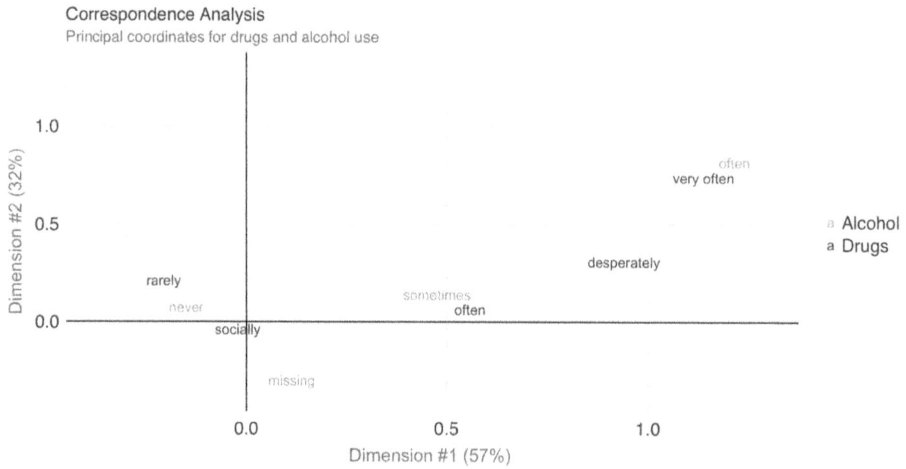

Figure 4-4. *Correspondence analysis principal coordinates for drug and alcohol use*

In Figure 4-4, we see that the correspondence analysis procedure has transformed the factors into variables called *principal coordinates*, which correspond to the axes in the plot and represent how much information in the contingency table they contain. We can, for example, interpret the proximity of "drinking often" and "using drugs very often" as indicating association.

This concludes our discussion on EDA. Let's proceed to feature engineering.

Feature Engineering

The feature engineering exercise comprises transforming the data to increase the performance of the model. This can include things like centering and scaling numerical values and performing string manipulation to extract meaningful variables. It also often includes variable selection—the process of selecting which predictors are used in the model.

In Figure 4-1 we see that the age variable has a range from 18 to over 60. Some algorithms, especially neural networks, train faster if we normalize our inputs so that they are of the same magnitude. Let's now normalize the age variable by removing the mean and scaling to unit variance, beginning by calculating its mean and standard deviation:

```
scale_values <- okc_train %>%
  summarise(
    mean_age = mean(age),
    sd_age = sd(age)
  ) %>%
  collect()

scale_values

# A tibble: 1 x 2
  mean_age sd_age
     <dbl>  <dbl>
1     32.3   9.44
```

We can then use these to transform the dataset:

```
okc_train <- okc_train %>%
  mutate(scaled_age = (age - !!scale_values$mean_age) /
           !!scale_values$sd_age)

dbplot_histogram(okc_train, scaled_age)
```

In Figure 4-5, we see that the scaled age variable has values that are closer to zero. We now move on to discussing other types of transformations, but during your feature engineering workflow you might want to perform the normalization for all numeric variables that you want to include in the model.

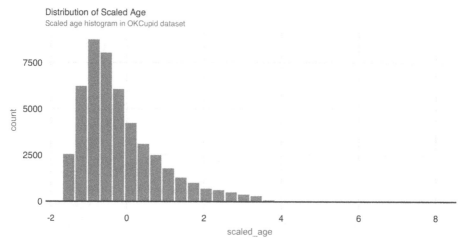

Figure 4-5. Distribution of scaled age

Since some of the profile features are multiple-select—in other words, a person can choose to associate multiple options for a variable—we need to process them before we can build meaningful models. If we take a look at the ethnicity column, for example, we see that there are many different combinations:

```
okc_train %>%
  group_by(ethnicity) %>%
  tally()
# Source: spark<?> [?? x 2]
   ethnicity                                      n
   <chr>                                      <dbl>
 1 hispanic / latin, white                     1051
 2 black, pacific islander, hispanic / latin      2
 3 asian, black, pacific islander                 5
 4 black, native american, white                 91
 5 middle eastern, white, other                  34
 6 asian, other                                  78
 7 asian, black, white                           12
 8 asian, hispanic / latin, white, other          7
 9 middle eastern, pacific islander               1
10 indian, hispanic / latin                       5
# … with more rows
```

One way to proceed would be to treat each combination of races as a separate level, but that would lead to a very large number of levels, which becomes problematic in many algorithms. To better encode this information, we can create dummy variables for each race, as follows:

```
ethnicities <- c("asian", "middle eastern", "black", "native american", "indian",
                 "pacific islander", "hispanic / latin", "white", "other")
ethnicity_vars <- ethnicities %>%
```

```
  purrr::map(~ expr(ifelse(like(ethnicity, !!.x), 1, 0))) %>%
  purrr::set_names(paste0("ethnicity_", gsub("\\s|/", "", ethnicities)))
okc_train <- mutate(okc_train, !!!ethnicity_vars)
okc_train %>%
  select(starts_with("ethnicity_")) %>%
  glimpse()

Observations: ??
Variables: 9
Database: spark_connection
$ ethnicity_asian          <dbl> 0, 0, 0, 0, 0, 0, 0, 0, 0, 0…
$ ethnicity_middleeastern  <dbl> 0, 0, 0, 0, 0, 0, 0, 0, 0, 0…
$ ethnicity_black          <dbl> 0, 1, 0, 0, 0, 0, 0, 0, 0, 0…
$ ethnicity_nativeamerican <dbl> 0, 0, 0, 0, 0, 0, 0, 0, 0, 0…
$ ethnicity_indian         <dbl> 0, 0, 0, 0, 0, 0, 0, 0, 0, 0…
$ ethnicity_pacificislander <dbl> 0, 0, 0, 0, 0, 0, 0, 0, 0, 0…
$ ethnicity_hispaniclatin  <dbl> 0, 0, 0, 0, 0, 0, 0, 0, 0, 0…
$ ethnicity_white          <dbl> 1, 0, 1, 0, 1, 1, 1, 0, 1, 0…
$ ethnicity_other          <dbl> 0, 0, 0, 0, 0, 0, 0, 0, 0, 0…
```

For the free text fields, a straightforward way to extract features is counting the total number of characters. We will store the train dataset in Spark's memory with com pute() to speed up computation.

```
okc_train <- okc_train %>%
  mutate(
    essay_length = char_length(paste(!!!syms(paste0("essay", 0:9))))
  ) %>% compute()

dbplot_histogram(okc_train, essay_length, bins = 100)
```

We can see the distribution of the essay_length variable in Figure 4-6.

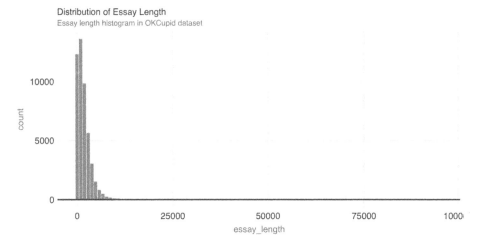

Figure 4-6. Distribution of essay length

We use this dataset in Chapter 5, so let's save it first as a Parquet file—an efficient file format ideal for numeric data:

```
spark_write_parquet(okc_train, "data/okc-train.parquet")
```

Now that we have a few more features to work with, we can begin running some unsupervised learning algorithms.

Supervised Learning

Once we have a good grasp on our dataset, we can start building some models. Before we do so, however, we need to come up with a plan to tune and validate the "candidate" models—in modeling projects, we often try different types of models and ways to fit them to see which ones perform the best. Since we are dealing with a binary classification problem, the metrics we can use include accuracy, precision, sensitivity, and area under the receiver operating characteristic curve (ROC AUC), among others. The metric you optimize depends on your specific business problem, but for this exercise, we will focus on the ROC AUC.

It is important that we don't peek at the testing holdout set until the very end, because any information we obtain could influence our modeling decisions, which would in turn make our estimates of model performance less credible. For tuning and validation, we perform 10-fold cross-validation, which is a standard approach for model tuning. The scheme works as follows: we first divide our dataset into 10 approximately equal-sized subsets. We take the 2nd to 10th sets together as the training set for an algorithm and validate the resulting model on the 1st set. Next, we reserve the 2nd set as the validation set and train the algorithm on the 1st and 3rd to 10th sets. In total, we train 10 models and average the performance. If time and resources allow, you can also perform this procedure multiple times with different random partitions of the data. In our case, we will demonstrate how to perform the cross-validation once. Hereinafter, we refer to the training set associated with each split as the *analysis* data, and the validation set as *assessment* data.

Using the `sdf_random_split()` function, we can create a list of subsets from our `okc_train` table:

```
vfolds <- sdf_random_split(
  okc_train,
  weights = purrr::set_names(rep(0.1, 10), paste0("fold", 1:10)),
  seed = 42
)
```

We then create our first analysis/assessment split as follows:

```
analysis_set <- do.call(rbind, vfolds[2:10])
assessment_set <- vfolds[[1]]
```

One item we need to carefully treat here is the scaling of variables. We need to make sure that we do not leak any information from the assessment set to the analysis set, so we calculate the mean and standard deviation on the analysis set only and apply the same transformation to both sets. Here is how we would handle this for the age variable:

```
make_scale_age <- function(analysis_data) {
  scale_values <- analysis_data %>%
    summarise(
      mean_age = mean(age),
      sd_age = sd(age)
    ) %>%
    collect()

  function(data) {
    data %>%
      mutate(scaled_age = (age - !!scale_values$mean_age) /
      !!scale_values$sd_age)
  }
}

scale_age <- make_scale_age(analysis_set)
train_set <- scale_age(analysis_set)
validation_set <- scale_age(assessment_set)
```

For brevity, here we show only how to transform the age variable. In practice, however, you would want to normalize each one of your continuous predictors, such as the essay_length variable we derived in the previous section.

Logistic regression is often a reasonable starting point for binary classification problems, so let's give it a try. Suppose also that our domain knowledge provides us with an initial set of predictors. We can then fit a model by using the Formula interface:

```
lr <- ml_logistic_regression(
  analysis_set, not_working ~ scaled_age + sex + drinks + drugs + essay_length
)
lr

Formula: not_working ~ scaled_age + sex + drinks + drugs + essay_length

Coefficients:
      (Intercept)        scaled_age               sex_m    drinks_socially
    -2.823517e+00     -1.309498e+00       -1.918137e-01       2.235833e-01
    drinks_rarely      drinks_often drinks_not at all    drinks_missing
     6.732361e-01      7.572970e-02        8.214072e-01      -4.456326e-01
drinks_very often       drugs_never      drugs_missing    drugs_sometimes
     8.032052e-02     -1.712702e-01       -3.995422e-01      -7.483491e-02
     essay_length
     3.664964e-05
```

To obtain a summary of performance metrics on the assessment set, we can use the ml_evaluate() function:

```
validation_summary <- ml_evaluate(lr, assessment_set)
```

You can print validation_summary to see the available metrics:

```
validation_summary

BinaryLogisticRegressionSummaryImpl
 Access the following via `$` or `ml_summary()`.
 - features_col()
 - label_col()
 - predictions()
 - probability_col()
 - area_under_roc()
 - f_measure_by_threshold()
 - pr()
 - precision_by_threshold()
 - recall_by_threshold()
 - roc()
 - prediction_col()
 - accuracy()
 - f_measure_by_label()
 - false_positive_rate_by_label()
 - labels()
 - precision_by_label()
 - recall_by_label()
 - true_positive_rate_by_label()
 - weighted_f_measure()
 - weighted_false_positive_rate()
 - weighted_precision()
 - weighted_recall()
 - weighted_true_positive_rate()
```

We can plot the ROC curve by collecting the output of validation_summary$roc() and using ggplot2:

```
roc <- validation_summary$roc() %>%
  collect()

ggplot(roc, aes(x = FPR, y = TPR)) +
  geom_line() + geom_abline(lty = "dashed")
```

Figure 4-7 shows the results of the plot.

The ROC curve plots the true positive rate (sensitivity) against the false positive rate (1–specificity) for varying values of the classification threshold. In practice, the business problem helps to determine where on the curve one sets the threshold for classification. The AUC is a summary measure for determining the quality of a model, and we can compute it by calling the area_under_roc() function.

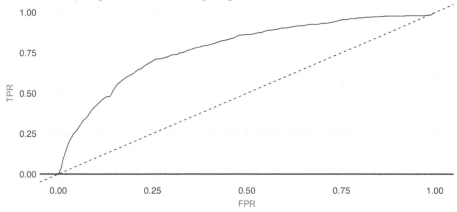

ROC Curve
Receiver operating characteristic curve for the logistic regression model

Figure 4-7. ROC curve for the logistic regression model

```
validation_summary$area_under_roc()
```

```
[1] 0.7872754
```

 Spark provides evaluation methods for only generalized
linear models (including linear models and logistic regression).
For other algorithms, you can use the evaluator functions (e.g.,
`ml_binary_classification_evaluator()` on the prediction Data-
Frame) or compute your own metrics.

Now, we can easily repeat the logic we already have and apply it to each analysis/
assessment split:

```
cv_results <- purrr::map_df(1:10, function(v) {
  analysis_set <- do.call(rbind, vfolds[setdiff(1:10, v)]) %>% compute()
  assessment_set <- vfolds[[v]]

  scale_age <- make_scale_age(analysis_set)
  train_set <- scale_age(analysis_set)
  validation_set <- scale_age(assessment_set)

  model <- ml_logistic_regression(
    analysis_set, not_working ~ scaled_age + sex + drinks + drugs + essay_length
  )
  s <- ml_evaluate(model, assessment_set)
  roc_df <- s$roc() %>%
    collect()
  auc <- s$area_under_roc()

  tibble(
    Resample = paste0("Fold", stringr::str_pad(v, width = 2, pad = "0")),
```

```
      roc_df = list(roc_df),
      auc = auc
    )
  })
```

This gives us 10 ROC curves:

```
unnest(cv_results, roc_df) %>%
  ggplot(aes(x = FPR, y = TPR, color = Resample)) +
  geom_line() + geom_abline(lty = "dashed")
```

Figure 4-8 shows the results of the plot.

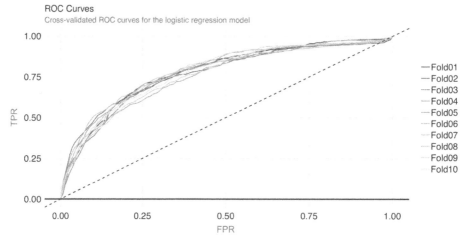

Figure 4-8. Cross-validated ROC curves for the logistic regression model

And we can obtain the average AUC metric:

```
mean(cv_results$auc)
```

```
[1] 0.7715102
```

Generalized Linear Regression

If you are interested in generalized linear model (GLM) diagnostics, you can also fit a logistic regression via the generalized linear regression interface by specifying `family = "binomial"`. Because the result is a regression model, the `ml_predict()` method does not give class probabilities. However, it includes confidence intervals for coefficient estimates:

```
glr <- ml_generalized_linear_regression(
  analysis_set,
  not_working ~ scaled_age + sex + drinks + drugs,
  family = "binomial"
)
```

```
tidy_glr <- tidy(glr)
```

We can extract the coefficient estimates into a tidy DataFrame, which we can then process further—for example, to create a coefficient plot, which you can see in Figure 4-9:

```
tidy_glr %>%
  ggplot(aes(x = term, y = estimate)) +
  geom_point() +
  geom_errorbar(
    aes(ymin = estimate - 1.96 * std.error,
        ymax = estimate + 1.96 * std.error, width = .1)
  ) +
  coord_flip() +
  geom_hline(yintercept = 0, linetype = "dashed")
```

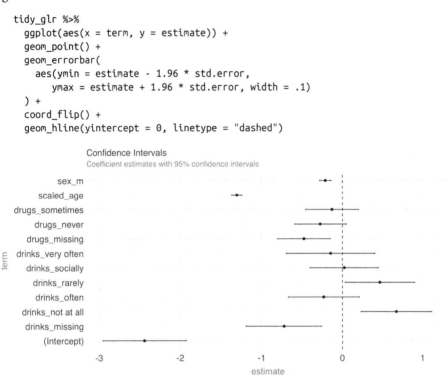

Figure 4-9. Coefficient estimates with 95% confidence intervals

 Both `ml_logistic_regression()` and `ml_linear_regression()` support elastic net regularization[5] through the `reg_param` and `elastic_net_param` parameters. `reg_param` corresponds to λ, whereas `elastic_net_param` corresponds to α. `ml_generalized_linear_regression()` supports only `reg_param`.

---

5 Zou H, Hastie T (2005). "Regularization and variable selection via the elastic net." *Journal of the royal statistical society: series B (statistical methodology)*, 67(2), 301–320.

Other Models

Spark supports many of the standard modeling algorithms and it's easy to apply these models and hyperparameters (values that control the model-fitting process) for your particular problem. You can find a list of supported ML-related functions in the appendix. The interfaces to access these functionalities are largely identical, so it is easy to experiment with them. For example, to fit a neural network model we can run the following:

```
nn <- ml_multilayer_perceptron_classifier(
  analysis_set,
  not_working ~ scaled_age + sex + drinks + drugs + essay_length,
  layers = c(12, 64, 64, 2)
)
```

This gives us a feedforward neural network model with two hidden layers of 64 nodes each. Note that you have to specify the correct values for the input and output layers in the `layers` argument. We can obtain predictions on a validation set using `ml_predict()`:

```
predictions <- ml_predict(nn, assessment_set)
```

Then, we can compute the AUC via `ml_binary_classification_evaluator()`:

```
ml_binary_classification_evaluator(predictions)

[1] 0.7812709
```

Up until now, we have not looked into the unstructured text in the essay fields apart from doing simple character counts. In the next section, we explore the textual data in more depth.

Unsupervised Learning

Along with speech, images, and videos, textual data is one of the components of the big data explosion. Prior to modern text-mining techniques and the computational resources to support them, companies had little use for freeform text fields. Today, text is considered a rich source of insights that can be found anywhere from physician's notes to customer complaints. In this section, we show some basic text analysis capabilities of `sparklyr`. If you would like more background on text-mining techniques, we recommend reading *Text Mining with R* (*https://oreil.ly/OrjWr*) by David Robinson and Julie Silge (O'Reilly).

In this section, we show how to perform a basic topic-modeling task on the essay data in the `OKCupid` dataset. Our plan is to concatenate the essay fields (of which there are 10) of each profile and regard each profile as a document, then attempt to discover *topics* (we define these soon) using Latent Dirichlet Allocation (LDA).

Data Preparation

As always, before analyzing a dataset (or a subset of one), we want to take a quick look at it to orient ourselves. In this case, we are interested in the freeform text that the users entered into their dating profiles.

```
essay_cols <- paste0("essay", 0:9)
essays <- okc %>%
  select(!!essay_cols)
essays %>%
  glimpse()

Observations: ??
Variables: 10
Database: spark_connection
$ essay0 <chr> "about me:<br />\n<br />\ni would love to think that…
$ essay1 <chr> "currently working as an international agent for a f…
$ essay2 <chr> "making people laugh.<br />\nranting about a good sa…
$ essay3 <chr> "the way i look. i am a six foot half asian, half ca…
$ essay4 <chr> "books:<br />\nabsurdistan, the republic, of mice an…
$ essay5 <chr> "food.<br />\nwater.<br />\ncell phone.<br />\nshelt…
$ essay6 <chr> "duality and humorous things", "missing", "missing",…
$ essay7 <chr> "trying to find someone to hang out with. i am down …
$ essay8 <chr> "i am new to california and looking for someone to w…
$ essay9 <chr> "you want to be swept off your feet!<br />\nyou are …
```

Just from this output, we see the following:

- The text contains HTML tags
- The text contains the newline (\n) character
- There are missing values in the data

The HTML tags and special characters pollute the data since they are not directly input by the user and do not provide interesting information. Similarly, since we have encoded missing character fields with the *missing* string, we need to remove it. (Note that by doing this we are also removing instances of the word "missing" written by the users, but the information lost from this removal is likely to be small.)

As you analyze your own text data, you will quickly come across and become familiar with the peculiarities of the specific dataset. As with tabular numerical data, preprocessing text data is an iterative process, and after a few tries we have the following transformations:

```
essays <- essays %>%
  # Replace `missing` with empty string.
  mutate_all(list(~ ifelse(. == "missing", "", .))) %>%
  # Concatenate the columns.
  mutate(essay = paste(!!!syms(essay_cols))) %>%
  # Remove miscellaneous characters and HTML tags.
  mutate(words = regexp_replace(essay, "\\n| |<[^>]*>|[^A-Za-z|']", " "))
```

Note here we are using `regex_replace()`, which is a Spark SQL function. Next, we discuss LDA and how to apply it to our cleaned dataset.

Topic Modeling

LDA is a type of topic model for identifying abstract "topics" in a set of documents. It is an unsupervised algorithm in that we do not provide any labels, or topics, for the input documents. LDA posits that each document is a mixture of topics, and each topic is a mixture of words. During training, it attempts to estimate both of these simultaneously. A typical use case for topic models involves categorizing many documents, for which the large number of documents renders manual approaches infeasible. The application domains range from GitHub issues to legal documents.

After we have a reasonably clean dataset following the workflow in the previous section, we can fit an LDA model with `ml_lda()`:

```
stop_words <- ml_default_stop_words(sc) %>%
  c(
    "like", "love", "good", "music", "friends", "people", "life",
    "time", "things", "food", "really", "also", "movies"
  )

lda_model <-  ml_lda(essays, ~ words, k = 6, max_iter = 1, min_token_length = 4,
                     stop_words = stop_words, min_df = 5)
```

We are also including a **stop_words** vector, consisting of commonly used English words and common words in our dataset, that instructs the algorithm to ignore them. After the model is fit, we can use the `tidy()` function to extract the associated betas, which are the per-topic-per-word probabilities, from the model.

```
betas <- tidy(lda_model)
betas

# A tibble: 256,992 x 3
   topic term       beta
   <int> <chr>      <dbl>
 1     0 know       303.
 2     0 work       250.
 3     0 want       367.
 4     0 books      211.
 5     0 family     213.
 6     0 think      291.
 7     0 going      160.
 8     0 anything   292.
 9     0 enjoy      145.
10     0 much       272.
# ... with 256,982 more rows
```

We can then visualize this output by looking at word probabilities by topic. In Figure 4-10 and Figure 4-11, we show the results at 1 iteration and 100 iterations. The code that generates Figure 4-10 follows; to generate Figure 4-11, you would need to set `max_iter = 100` when running `ml_lda()`, but beware that this can take a really long time in a single machine—this is the kind of big-compute problem that a proper Spark cluster would be able to easily tackle.

```
betas %>%
  group_by(topic) %>%
  top_n(10, beta) %>%
  ungroup() %>%
  arrange(topic, -beta) %>%
  mutate(term = reorder(term, beta)) %>%
  ggplot(aes(term, beta, fill = factor(topic))) +
    geom_col(show.legend = FALSE) +
    facet_wrap(~ topic, scales = "free") +
    coord_flip()
```

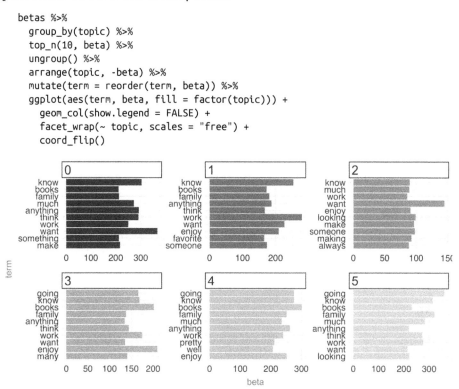

Figure 4-10. The most common terms per topic in the first iteration

At 100 iterations, we can see "topics" starting to emerge. This could be interesting information in its own right if you were digging into a large collection of documents with which you aren't familiar. The learned topics can also serve as features in a downstream supervised learning task; for example, we could consider using the topic number as a predictor in our model to predict employment status in our predictive modeling example.

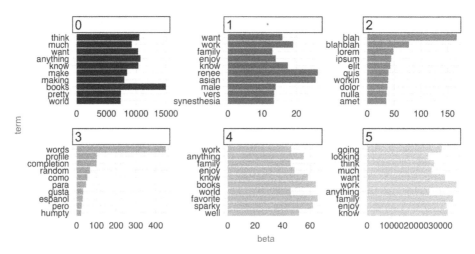

Figure 4-11. The most common terms per topic after 100 iterations

Finally, to conclude this chapter you should disconnect from Spark. Chapter 5 also makes use of the OKCupid dataset, but we provide instructions to reload it from scratch:

```
spark_disconnect(sc)
```

Recap

In this chapter, we covered the basics of building predictive models in Spark with R by presenting the topics of EDA, feature engineering, and building supervised models, in which we explored using logistic regression and neural networks—just to pick a few from dozens of models available in Spark through MLlib.

We then explored how to use unsupervised learning to process raw text, in which you created a topic model that automatically grouped the profiles into six categories. We demonstrated that building the topic model can take a significant amount of time using a single machine, which is a nearly perfect segue to introduce full-sized computing clusters! But hold that thought: we first need to consider how to automate data science workflows.

As we mentioned when introducing this chapter, emphasis was placed on predictive modeling. Spark can help with data science at scale, but it can also assist in productionizing data science workflows into automated processes, known by many as machine learning. Chapter 5 presents the tools we will need to take our predictive models, and even our entire training workflows, into automated environments that can run continuously or be exported and consumed in web applications, mobile applications, and more.

Pipelines

You will never walk again, but you will fly!
—*Three-Eyed Raven*

In Chapter 4, you learned how to build predictive models using the high-level functions Spark provides and well-known R packages that work well together with Spark. You learned about supervised methods first and finished the chapter with an unsupervised method over raw text.

In this chapter, we dive into Spark Pipelines, which is the engine that powers the features we demonstrated in Chapter 4. So, for instance, when you invoke an MLlib function via the formula interface in R—for example, ml_logistic_regres sion(cars, am ~ .)—a *pipeline* is constructed for you under the hood. Therefore, Pipelines also allow you to make use of advanced data processing and modeling workflows. In addition, a pipeline also facilitates collaboration across data science and engineering teams by allowing you to *deploy* pipelines into production systems, web applications, mobile applications, and so on.

This chapter also happens to be the last chapter that encourages using your local computer as a Spark cluster. You are just one chapter away from getting properly introduced to cluster computing and beginning to perform data science or machine learning that can scale to the most demanding computation problems.

Overview

The building blocks of pipelines are objects called *transformers* and *estimators*, which are collectively referred to as *pipeline stages*. A *transformer* can be used to apply transformations to a DataFrame and return another DataFrame; the resulting DataFrame often comprises the original DataFrame with new columns appended to it. An *estimator*, on the other hand, can be used to create a transformer giving some training data. Consider the following example to illustrate this relationship: a "center and scale" estimator can learn the mean and standard deviation of some data and store the statistics in a resulting transformer object; this transformer can then be used to normalize the data that it was trained on and also any new, yet unseen, data.

Here is an example of how to define an estimator:

```
library(sparklyr)
library(dplyr)

sc <- spark_connect(master = "local", version = "2.3")

scaler <- ft_standard_scaler(
  sc,
  input_col = "features",
  output_col = "features_scaled",
  with_mean = TRUE)

scaler

StandardScaler (Estimator)
<standard_scaler_7f6d46f452a1>
 (Parameters -- Column Names)
  input_col: features
  output_col: features_scaled
 (Parameters)
  with_mean: TRUE
  with_std: TRUE
```

We can now create some data (for which we know the mean and standard deviation) and then fit our scaling model to it using the ml_fit() function:

```
df <- copy_to(sc, data.frame(value = rnorm(100000))) %>%
  ft_vector_assembler(input_cols = "value", output_col = "features")

scaler_model <- ml_fit(scaler, df)
scaler_model

StandardScalerModel (Transformer)
<standard_scaler_7f6d46f452a1>
 (Parameters -- Column Names)
  input_col: features
  output_col: features_scaled
 (Transformer Info)
  mean:  num 0.00421
  std:   num 0.999
```

 In Spark ML, many algorithms and feature transformers require that the input be a vector column. The function ft_vector_assem bler() performs this task. You can also use the function to initialize a transformer to be used in a pipeline.

We see that the mean and standard deviation are very close to 0 and 1, respectively, which is what we expect. We then can use the transformer to *transform* a DataFrame, using the ml_transform() function:

```
scaler_model %>%
  ml_transform(df) %>%
  glimpse()

Observations: ??
Variables: 3
Database: spark_connection
$ value          <dbl> 0.75373300, -0.84207731, 0.59365113, -…
$ features       <list> [0.753733, -0.8420773, 0.5936511, -0.…
$ features_scaled <list> [0.7502211, -0.8470762, 0.58999, -0.4…
```

Now that you've seen basic examples of estimators and transformers, we can move on to pipelines.

Creation

A *pipeline* is simply a sequence of transformers and estimators, and a *pipeline model* is a pipeline that has been trained on data so all of its components have been converted to transformers.

There are a couple of ways to construct a pipeline in sparklyr, both of which use the ml_pipeline() function.

We can initialize an empty pipeline with ml_pipeline(sc) and append stages to it:

```
ml_pipeline(sc) %>%
  ft_standard_scaler(
    input_col = "features",
    output_col = "features_scaled",
    with_mean = TRUE)

Pipeline (Estimator) with 1 stage
<pipeline_7f6d6a6a38ee>
  Stages
  |--1 StandardScaler (Estimator)
  |     <standard_scaler_7f6d63bfc7d6>
  |     (Parameters -- Column Names)
  |      input_col: features
  |      output_col: features_scaled
  |     (Parameters)
  |      with_mean: TRUE
  |      with_std: TRUE
```

Alternatively, we can pass stages directly to ml_pipeline():

```
pipeline <- ml_pipeline(scaler)
```

We fit a pipeline as we would fit an estimator:

```
pipeline_model <- ml_fit(pipeline, df)
pipeline_model

PipelineModel (Transformer) with 1 stage
<pipeline_7f6d64df6e45>
  Stages
  |--1 StandardScalerModel (Transformer)
  |     <standard_scaler_7f6d46f452a1>
  |     (Parameters -- Column Names)
  |      input_col: features
  |      output_col: features_scaled
  |     (Transformer Info)
  |      mean:  num 0.00421
  |      std:   num 0.999

pipeline
```

 As a result of the design of Spark ML, pipelines are always estimator objects, even if they comprise only transformers. This means
that if you have a pipeline with only transformers, you still need to
call ml_fit() on it to obtain a transformer. The "fitting" procedure
in this case wouldn't actually modify any of the transformers.

Use Cases

Now that you have an understanding of the rudimentary concepts for ML Pipelines,
let's apply them to the predictive modeling problem from the previous chapter in
which we are trying to predict whether people are currently employed by looking at
their profiles. Our starting point is the okc_train DataFrame with the relevant
columns.

```
okc_train <- spark_read_parquet(sc, "data/okc-train.parquet")

okc_train <- okc_train %>%
  select(not_working, age, sex, drinks, drugs, essay1:essay9, essay_length)
```

We first exhibit the pipeline, which includes feature engineering and modeling steps,
and then walk through it:

```
pipeline <- ml_pipeline(sc) %>%
  ft_string_indexer(input_col = "sex", output_col = "sex_indexed") %>%
  ft_string_indexer(input_col = "drinks", output_col = "drinks_indexed") %>%
  ft_string_indexer(input_col = "drugs", output_col = "drugs_indexed") %>%
  ft_one_hot_encoder_estimator(
    input_cols = c("sex_indexed", "drinks_indexed", "drugs_indexed"),
    output_cols = c("sex_encoded", "drinks_encoded", "drugs_encoded")
  ) %>%
  ft_vector_assembler(
    input_cols = c("age", "sex_encoded", "drinks_encoded",
                   "drugs_encoded", "essay_length"),
    output_col = "features"
```

```
) %>%
ft_standard_scaler(input_col = "features", output_col = "features_scaled",
                   with_mean = TRUE) %>%
ml_logistic_regression(features_col = "features_scaled",
                       label_col = "not_working")
```

The first three stages index the sex, drinks, and drugs columns, which are characters, into numeric indices via ft_string_indexer(). This is necessary for the
ft_one_hot_encoder_estimator() that comes next, which requires numeric column
inputs. When all of our predictor variables are of numeric type (recall that age is
numeric already), we can create our features vector using ft_vector_assembler(),
which concatenates all of its inputs together into one column of vectors. We can then
use ft_standard_scaler() to normalize all elements of the features column (including the one-hot-encoded 0/1 values of the categorical variables), and finally apply a
logistic regression via ml_logistic_regression().

During prototyping, you might want to execute these transformations *eagerly* on a
small subset of the data, by passing the DataFrame to the ft_ and ml_ functions, and
inspecting the transformed DataFrame. The immediate feedback allows for rapid iteration of ideas; when you have arrived at the desired processing steps, you can roll
them up into a pipeline. For example, you can do the following:

```
okc_train %>%
  ft_string_indexer("sex", "sex_indexed") %>%
  select(sex_indexed)

# Source: spark<?> [?? x 1]
   sex_indexed
         <dbl>
 1           0
 2           0
 3           1
 4           0
 5           1
 6           0
 7           0
 8           1
 9           1
10           0
# … with more rows
```

After you have found the appropriate transformations for your dataset, you can
replace the DataFrame input with ml_pipeline(sc), and the result will be a pipeline
that you can apply to any DataFrame with the appropriate schema. In the next section, we'll see how pipelines can make it easier for us to test different model
specifications.

Hyperparameter Tuning

Going back to the pipeline we have created earlier, we can use `ml_cross_valida tor()` to perform the cross-validation workflow we demonstrated in the previous chapter and easily test different hyperparameter combinations. In this example, we test whether centering the variables improves predictions together with various regularization values for the logistic regression. We define the cross-validator as follows:

```
cv <- ml_cross_validator(
  sc,
  estimator = pipeline,
  estimator_param_maps = list(
    standard_scaler = list(with_mean = c(TRUE, FALSE)),
    logistic_regression = list(
      elastic_net_param = c(0.25, 0.75),
      reg_param = c(1e-2, 1e-3)
    )
  ),
  evaluator = ml_binary_classification_evaluator(sc, label_col = "not_working"),
  num_folds = 10)
```

The `estimator` argument is simply the estimator that we want to tune, and in this case it is the `pipeline` that we defined. We provide the hyperparameter values we are interested in via the `estimator_param_maps` parameter, which takes a nested named list. The names at the first level correspond to UIDs, which are unique identifiers associated with each pipeline stage object, of the stages we want to tune (if a partial UID is provided, `sparklyr` will attempt to match it to a pipeline stage), and the names at the second level correspond to parameters of each stage. In the preceding snippet, we are specifying that we want to test the following:

Standard scaler
    The values `TRUE` and `FALSE` for `with_mean`, which denotes whether predictor values are centered.

Logistic regression
    The values `0.25` and `0.75` for α, and the values `1e-2` and `1e-3` for λ.

We expect this to give rise to $2 \times 2 \times 2 = 8$ hyperparameter combinations, which we can confirm by printing the cv object:

```
cv

CrossValidator (Estimator)
<cross_validator_d5676ac6f5>
 (Parameters -- Tuning)
  estimator: Pipeline
             <pipeline_d563b0cba31>
  evaluator: BinaryClassificationEvaluator
             <binary_classification_evaluator_d561d90b53d>
    with metric areaUnderROC
```

```
num_folds: 10
[Tuned over 8 hyperparameter sets]
```

As with any other estimator, we can fit the cross-validator by using `ml_fit()`

```
cv_model <- ml_fit(cv, okc_train)
```

and then inspect the results:

```
ml_validation_metrics(cv_model) %>%
  arrange(-areaUnderROC)

  areaUnderROC elastic_net_param_1 reg_param_1 with_mean_2
1    0.7722700                0.75       0.001        TRUE
2    0.7718431                0.75       0.010       FALSE
3    0.7718350                0.75       0.010        TRUE
4    0.7717677                0.25       0.001        TRUE
5    0.7716070                0.25       0.010        TRUE
6    0.7715972                0.25       0.010       FALSE
7    0.7713816                0.75       0.001       FALSE
8    0.7703913                0.25       0.001       FALSE
```

Now that we have seen the pipelines API in action, let's talk more formally about how they behave in various contexts.

Operating Modes

By now, you have likely noticed that the pipeline stage functions, such as `ft_string_indexer()` and `ml_logistic_regression()`, return different types of objects depending on the first argument passed to them. Table 5-1 presents the full pattern.

Table 5-1. Operating modes in machine learning functions

First argument	Returns	Example
Spark connection	Estimator or transformer object	`ft_string_indexer(sc)`
Pipeline	Pipeline	`ml_pipeline(sc) %>% ft_string_indexer()`
DataFrame, without formula	DataFrame	`ft_string_indexer(iris, "Species", "indexed")`
DataFrame, with formula	sparklyr ML model object	`ml_logistic_regression(iris, Species ~ .)`

These functions are implemented using S3 (*https://adv-r.hadley.nz/s3.html*), which is the most popular object-oriented programming paradigm provided by R. For our purposes, it suffices to know that the behavior of an `ml_` or `ft_` function is dictated by the class of the first argument provided. This allows us to provide a wide range of features without introducing additional function names. We now can summarize the behavior of these functions:

- If a Spark connection is provided, the function returns a transformer or estimator object, which can be utilized directly using ml_fit() or ml_transform() or be included in a pipeline.

- If a pipeline is provided, the function returns a pipeline object with the stage appended to it.

- If a DataFrame is provided to a feature transformer function (those with prefix ft_), or an ML algorithm without also providing a formula, the function instantiates the pipeline stage object, fits it to the data if necessary (if the stage is an estimator), and then transforms the DataFrame returning a DataFrame.

- If a DataFrame and a formula are provided to an ML algorithm that supports the formula interface, sparklyr builds a pipeline model under the hood and returns an ML model object that contains additional metadata information.

The formula interface approach is what we studied in Chapter 4, and this is what we recommend users new to Spark start with, since its syntax is similar to existing R modeling packages and abstracts away some Spark ML peculiarities. However, to take advantage of the full power of Spark ML and leverage pipelines for workflow organization and interoperability, it is worthwhile to learn the ML Pipelines API.

With the basics of pipelines down, we are now ready to discuss the collaboration and model deployment aspects hinted at in the introduction of this chapter.

Interoperability

One of the most powerful aspects of pipelines is that they can be serialized to disk and are fully interoperable with the other Spark APIs such as Python and Scala. This means that you can easily share them among users of Spark working in different languages, which might include other data scientists, data engineers, and deployment engineers. To save a pipeline model, call ml_save() and provide a path:

```
model_dir <- file.path("spark_model")
ml_save(cv_model$best_model, model_dir, overwrite = TRUE)

Model successfully saved.
```

Let's take a look at the directory to which we just wrote:

```
list.dirs(model_dir,full.names = FALSE) %>%
  head(10)

[1] ""
[2] "metadata"
[3] "stages"
[4] "stages/0_string_indexer_5b42c72817b"
[5] "stages/0_string_indexer_5b42c72817b/data"
[6] "stages/0_string_indexer_5b42c72817b/metadata"
```

```
  [7] "stages/1_string_indexer_5b423192b89f"
  [8] "stages/1_string_indexer_5b423192b89f/data"
  [9] "stages/1_string_indexer_5b423192b89f/metadata"
 [10] "stages/2_string_indexer_5b421796e826"
```

We can dive into a couple of the files to see what type of data was saved:

```
spark_read_json(sc, file.path(
  file.path(dir(file.path(model_dir, "stages"),
                pattern = "1_string_indexer.*",
                full.names = TRUE), "metadata")
)) %>%
  glimpse()

Observations: ??
Variables: 5
Database: spark_connection
$ class       <chr> "org.apache.spark.ml.feature.StringIndexerModel"
$ paramMap    <list> [["error", "drinks", "drinks_indexed", "frequencyDesc"]]
$ sparkVersion <chr> "2.3.2"
$ timestamp   <dbl> 1.561763e+12
$ uid         <chr> "string_indexer_ce05afa9899"

spark_read_parquet(sc, file.path(
  file.path(dir(file.path(model_dir, "stages"),
                pattern = "6_logistic_regression.*",
                full.names = TRUE), "data")
))

# Source: spark<data> [?? x 5]
  numClasses numFeatures interceptVector coefficientMatr… isMultinomial
       <int>       <int> <list>          <list>           <lgl>
1          2          12 <dbl [1]>       <-1.27950828662… FALSE
```

We see that quite a bit of information has been exported, from the SQL statement in the dplyr transformer to the fitted coefficient estimates of the logistic regression. We can then (in a new Spark session) reconstruct the model by using ml_load():

```
model_reload <- ml_load(sc, model_dir)
```

Let's see if we can retrieve the logistic regression stage from this pipeline model:

```
ml_stage(model_reload, "logistic_regression")

LogisticRegressionModel (Transformer)
<logistic_regression_5b423b539d0f>
 (Parameters -- Column Names)
  features_col: features_scaled
  label_col: not_working
  prediction_col: prediction
  probability_col: probability
  raw_prediction_col: rawPrediction
 (Transformer Info)
  coefficient_matrix:  num [1, 1:12] -1.2795 -0.0915 0 0.126 -0.0324 ...
  coefficients:  num [1:12] -1.2795 -0.0915 0 0.126 -0.0324 ...
```

```
intercept:  num -2.79
intercept_vector:  num -2.79
num_classes:  int 2
num_features:  int 12
threshold:  num 0.5
thresholds:  num [1:2] 0.5 0.5
```

Note that the exported JSON and parquet files are agnostic of the API that exported them. This means that in a multilingual machine learning engineering team, you can pick up a data preprocessing pipeline from a data engineer working in Python, build a prediction model on top of it, and then hand off the final pipeline to a deployment engineering working in Scala. In the next section, we discuss deployment of models in more detail.

 When ml_save() is called for sparklyr ML models (created using the formula interface), the associated pipeline model is saved, but any sparklyr-specific metadata, such as index labels, is not. In other words, saving a sparklyr ml_model object and then loading it will yield a pipeline model object, as if you created it via the ML Pipelines API. This behavior is required to use pipelines with other programming languages.

Before we move on to discuss how to run pipelines in production, make sure you disconnect from Spark:

```
spark_disconnect(sc)
```

That way, we can start from a brand new environment, which is also expected when you are deploying pipelines to production.

Deployment

What we have just demonstrated bears emphasizing: by collaborating within the framework of ML Pipelines, we reduce friction among different personas in a data science team. In particular, we cut down on the time from modeling to deployment.

In many cases, a data science project does not end with just a slide deck with insights and recommendations. Instead, the business problem at hand might require scoring new data points on a schedule or on-demand in real time. For example, a bank might want to evaluate its mortgage portfolio risk nightly or provide instant decisions on credit card applications. This process of taking a model and turning it into a service that others can consume is usually referred to as *deployment* or *productionization*. Historically, there was a large gap between the analyst who built the model and the engineer who deployed it: the former might work in R and develop extensive documentation on the scoring mechanism, so that the latter can reimplement the model in

C++ or Java. This practice, which might easily take months in some organizations, is less prevalent today, but is almost always unnecessary in Spark ML workflows.

The aforementioned nightly portfolio risk and credit application scoring examples represent two modes of ML deployment known as *batch* and *real time*. Loosely, batch processing implies processing many records at the same time, and that execution time is not important as long it is reasonable (often on the scale of minutes to hours). On the other hand, real-time processing implies scoring one or a few records at a time, but the latency is crucial (on the scale of <1 second). Let's turn now to see how we can take our OKCupid pipeline model to "production."

Batch Scoring

For both scoring methods, batch and real time, we will expose our model as web services, in the form of an API over the Hypertext Transfer Protocol (HTTP). This is the primary medium over which software communicates. By providing an API, other services or end users can utilize our model without any knowledge of R or Spark. The plumber (*https://www.rplumber.io/*) R package enables us to do this very easily by annotating our prediction function.

You will need make sure that plumber, callr, and the httr package are installed by running the following:

```
install.packages(c("plumber", "callr", "httr"))
```

The callr package provides support to run R code in separate R sessions; it is not strictly required, but we will use it to start a web service in the background. The httr package allows us to use web APIs from R.

In the batch scoring use case, we simply initiate a Spark connection and load the saved model. Save the following script as *plumber/spark-plumber.R*:

```
library(sparklyr)
sc <- spark_connect(master = "local", version = "2.3")

spark_model <- ml_load(sc, "spark_model")

#* @post /predict
score_spark <- function(age, sex, drinks, drugs, essay_length) {
  new_data <- data.frame(
    age = age,
    sex = sex,
    drinks = drinks,
    drugs = drugs,
    essay_length = essay_length,
    stringsAsFactors = FALSE
  )
  new_data_tbl <- copy_to(sc, new_data, overwrite = TRUE)
```

```
    ml_transform(spark_model, new_data_tbl) %>%
      dplyr::pull(prediction)
}
```

We can then initialize the service by executing the following:

```
service <- callr::r_bg(function() {
  p <- plumber::plumb("plumber/spark-plumber.R")
  p$run(port = 8000)
})
```

This starts the web service locally, and then we can query the service with new data to be scored; however, you might need to wait a few seconds for the Spark service to initialize:

```
httr::content(httr::POST(
  "http://127.0.0.1:8000/predict",
  body = '{"age": 42, "sex": "m", "drinks": "not at all",
          "drugs": "never", "essay_length": 99}'
))
[[1]]
[1] 0
```

This reply informs us that this particular profile is likely to not be unemployed—that is, employed. We can now terminate the plumber service by stopping the callr service:

```
service$interrupt()
```

If we were to time this operation (e.g., with system.time()), we see that the latency is on the order of hundreds of milliseconds, which might be appropriate for batch applications but is insufficient for real time. The main bottleneck is the serialization of the R DataFrame to a Spark DataFrame and back. Also, it requires an active Spark session, which is a heavy runtime requirement. To ameliorate these issues, we discuss next a deployment method more suitable for real-time deployment.

Real-Time Scoring

For real-time production, we want to keep dependencies as light as possible so we can target more platforms for deployment. We now show how we can use the mleap (*http://bit.ly/2Z7jgSV*) package, which provides an interface to the MLeap (*http://bit.ly/33G271R*) library, to serialize and serve Spark ML models. MLeap is open source (Apache License 2.0) and supports a wide range of, though not all, Spark ML transformers. At runtime, the only prerequisites for the environment are the Java Virtual Machine (JVM) and the MLeap runtime library. This avoids both the Spark binaries and expensive overhead in converting data to and from Spark DataFrames.

Since mleap is a sparklyr extension and an R package, first we need to install it from CRAN:

```
install.packages("mleap")
```

It then must be loaded when spark_connect() is called; so let's restart your R session, establish a new Spark connection,[1] and load the pipeline model that we previously saved:

```
library(sparklyr)
library(mleap)

sc <- spark_connect(master = "local", version = "2.3")

spark_model <- ml_load(sc, "spark_model")
```

The way we save a model to MLeap bundle format is very similar to saving a model using the Spark ML Pipelines API; the only additional argument is sample_input, which is a Spark DataFrame with schema that we expect new data to be scored to have:

```
sample_input <- data.frame(
  sex = "m",
  drinks = "not at all",
  drugs = "never",
  essay_length = 99,
  age = 25,
  stringsAsFactors = FALSE
)

sample_input_tbl <- copy_to(sc, sample_input)

ml_write_bundle(spark_model, sample_input_tbl, "mleap_model.zip", overwrite =
TRUE)
```

We can now deploy the artifact we just created, *mleap_model.zip*, in any device that runs Java and has the open source MLeap runtime dependencies, without needing Spark or R! In fact, we can go ahead and disconnect from Spark already:

```
spark_disconnect(sc)
```

Before we use this MLeap model, make sure the runtime dependencies are installed:

```
mleap::install_maven()
mleap::install_mleap()
```

To test this model, we can create a new plumber API to expose it. The script *plumber/ mleap-plumber.R* is very similar to the previous example:

```
library(mleap)

mleap_model <- mleap_load_bundle("mleap_model.zip")
```

---

1 As of the writing of this book, MLeap does not support Spark 2.4.

```
#* @post /predict
score_spark <- function(age, sex, drinks, drugs, essay_length) {
  new_data <- data.frame(
    age = as.double(age),
    sex = sex,
    drinks = drinks,
    drugs = drugs,
    essay_length = as.double(essay_length),
    stringsAsFactors = FALSE
  )
  mleap_transform(mleap_model, new_data)$prediction
}
```

And the way we launch the service is exactly the same:

```
service <- callr::r_bg(function() {
  p <- plumber::plumb("plumber/mleap-plumber.R")
  p$run(port = 8000)
})
```

We can run the exact same code we did previously to test unemployment predictions in this new service:

```
httr::POST(
  "http://127.0.0.1:8000/predict",
  body = '{"age": 42, "sex": "m", "drinks": "not at all",
          "drugs": "never", "essay_length": 99}'
) %>%
  httr::content()

[[1]]
[1] 0
```

If we were to time this operation, we would see that the service now returns predictions in tens of milliseconds.

Let's stop this service and then wrap up this chapter:

```
service$interrupt()
```

Recap

In this chapter, we discussed Spark Pipelines, which is the engine behind the modeling functions introduced in Chapter 4. You learned how to tidy up your predictive modeling workflows by organizing data processing and modeling algorithms into pipelines. You learned that pipelines also facilitate collaboration among members of a multilingual data science and engineering team by sharing a language-agnostic serialization format—you can export a Spark pipeline from R and let others reload your pipeline into Spark using Python or Scala, which allows them to collaborate without changing their language of choice.

You also learned how to deploy pipelines using `mleap`, a Java runtime that provides another path to productionize Spark models—you can export a pipeline and integrate it to Java-enabled environments without requiring the target environment to support Spark or R.

You probably noticed that some algorithms, especially the unsupervised learning kind, were slow, even for the `OKCupid` dataset, which can be loaded into memory. If we had access to a proper Spark cluster, we could spend more time modeling and less time waiting! Not only that, but we could use cluster resources to run broader hyperparameter-tuning jobs and process large datasets. To get there, Chapter 6 presents what exactly a computing cluster is and explains the various options you can consider, like building your own or using cloud clusters on demand.

## CHAPTER 6

Clusters

I have a very large army and very large dragons.
—Daenerys Targaryen

Previous chapters focused on using Spark over a single computing instance, your personal computer. In this chapter, we introduce techniques to run Spark over multiple computing instances, also known as a *computing* cluster. This chapter and subsequent ones will introduce and make use of concepts applicable to computing clusters; however, it's not required to use a computing cluster to follow along, so you can still use your personal computer. It's worth mentioning that while previous chapters focused on single computing instances, you can also use all the data analysis and modeling techniques we presented in a computing cluster without changing any code.

If you already have a Spark cluster in your organization, you could consider skipping to Chapter 7, which teaches you how to connect to an existing cluster. Otherwise, if you don't have a cluster or are considering improvements to your existing infrastructure, this chapter introduces the cluster trends, managers, and providers available today.

Overview

There are three major trends in cluster computing worth discussing: *on-premises*, *cloud* computing, and *Kubernetes*. Framing these trends over time will help us understand how they came to be, what they are, and what their future might be. To illustrate this, Figure 6-1 plots these trends over time using data from Google trends.

For on-premises clusters, you or someone in your organization purchased physical computers that were intended to be used for cluster computing. The computers in this cluster are made of *off-the-shelf* hardware, meaning that someone placed an order to purchase computers usually found on store shelves, or *high-performance* hardware, meaning that a computing vendor provided highly customized computing hardware, which also comes optimized for high-performance network connectivity, power consumption, and so on.

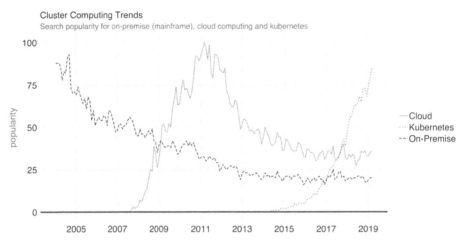

Figure 6-1. Google trends for on-premises (mainframe), cloud computing, and Kubernetes

When purchasing hundreds or thousands of computing instances, it doesn't make sense to keep them in the usual computing case that we are all familiar with; instead, it makes sense to stack them as efficiently as possible on top of one another to minimize the space the use. This group of efficiently stacked computing instances is known as a *rack* (*https://oreil.ly/zKOr-*). After a cluster grows to thousands of computers, you will also need to host hundreds of racks of computing devices; at this scale, you would also need significant physical space to host those racks.

A building that provides racks of computing instances is usually known as a *datacenter*. At the scale of a datacenter, you would also need to find ways to make the building more efficient, especially the cooling system, power supplies, network connectivity, and so on. Since this is time-consuming, a few organizations have come together to open source their infrastructure under the Open Compute Project (*http://www.opencompute.org/*) initiative, which provides a set of datacenter blueprints free for anyone to use.

There is nothing preventing you from building our own datacenter, and, in fact, many organizations have followed this path. For instance, Amazon started as an online bookstore, but over the years it grew to sell much more than just books. Along with its online store growth, its datacenters also grew in size. In 2002, Amazon considered renting servers in their datacenters to the public (*https://oreil.ly/Nx3BD*), and two years later, Amazon Web Services (AWS) launched as a way to let anyone rent servers in the company's datacenters on demand, meaning that you did not need to purchase, configure, maintain, or tear down your own clusters; rather, you could rent them directly from AWS.

This on-demand compute model is what we know today as *cloud computing*. In the cloud, the cluster you use is not owned by you, and it's not in your physical building; instead it's a datacenter owned and managed by someone else. Today, there are many cloud providers in this space, including AWS, Databricks, Google, Microsoft, Qubole, and many others. Most cloud computing platforms provide a user interface through either a web application or command line to request and manage resources.

While the benefits of processing data in the *cloud* were obvious for many years, picking a cloud provider had the unintended side effect of locking in organizations with one particular provider, making it hard to switch between providers or back to on-premises clusters. *Kubernetes*, announced by Google in 2014, is an open source system for managing containerized applications across multiple hosts (*https://oreil.ly/u6H5X*). In practice, it makes it easier to deploy across multiple cloud providers and on-premises as well.

In summary, we have seen a transition from on-premises to cloud computing and, more recently, Kubernetes. These technologies are often loosely described as the *private cloud*, the *public cloud*, and as one of the orchestration services that can enable a *hybrid cloud*, respectively. This chapter walks you through each cluster computing trend in the context of Spark and R.

On-Premises

As mentioned in the overview section, on-premises clusters represent a set of computing instances procured and managed by staff members from your organization. These clusters can be highly customized and controlled; however, they can also incur higher initial expenses and maintenance costs.

When using on-premises Spark clusters, there are two concepts you should consider:

Cluster manager
> In a similar way as to how an operating system (like Windows or macOS) allows you to run multiple applications in the same computer, a cluster manager allows multiple applications to be run in the same cluster. You need to choose one yourself when working with on-premises clusters.

Spark distribution
> While you can install Spark from the Apache Spark site, many organizations partner with companies that can provide support and enhancements to Apache Spark, which we often refer to as Spark *distributions*.

Managers

To run Spark within a computing cluster, you will need to run software capable of initializing Spark over each physical machine and register all the available computing nodes. This software is known as a cluster manager (*https://oreil.ly/Ye4zH*). The available cluster managers in Spark are *Spark Standalone*, *YARN*, *Mesos*, and *Kubernetes*.

> In distributed systems and clusters literature, we often refer to each physical machine as a *compute instance*, *compute node*, *instance*, or *node*.

Standalone

In *Spark Standalone*, Spark uses itself as its own cluster manager, which allows you to use Spark without installing additional software in your cluster. This can be useful if you are planning to use your cluster to run only Spark applications; if this cluster is not dedicated to Spark, a generic cluster manager like YARN, Mesos, or Kubernetes would be more suitable. The Spark Standalone documentation (*http://bit.ly/307YtM6*) contains detailed information on configuring, launching, monitoring, and enabling high availability, as illustrated in Figure 6-2.

However, since Spark Standalone is contained within a Spark installation, by completing Chapter 2, you have now a Spark installation available that you can use to initialize a local Spark Standalone cluster on your own machine. In practice, you would want to start the worker nodes on different machines, but for simplicity, we present the code to start a standalone cluster on a single machine.

First, retrieve the SPARK_HOME directory by running **spark_home_dir()**, and then start the master node and a worker node as follows:

```
# Retrieve the Spark installation directory
spark_home <- spark_home_dir()

# Build paths and classes
spark_path <- file.path(spark_home, "bin", "spark-class")

# Start cluster manager master node
system2(spark_path, "org.apache.spark.deploy.master.Master", wait = FALSE)
```

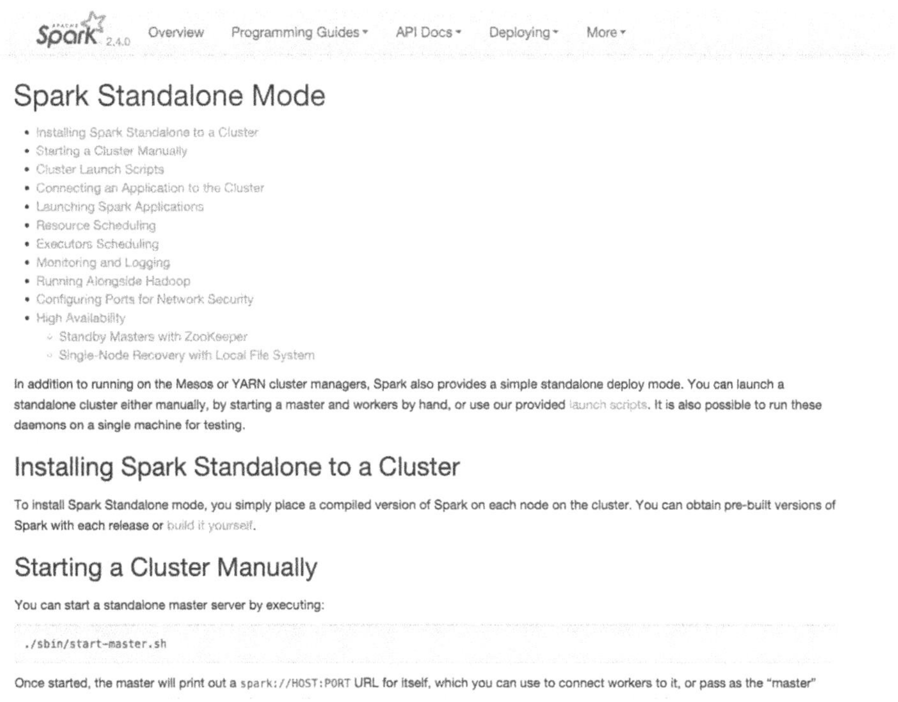

Spark 2.4.0
Overview
Programming Guides
API Docs
Deploying
More
Spark Standalone Mode
Installing Spark Standalone to a Cluster
Starting a Cluster Manually
Cluster Launch Scripts
Connecting an Application to the Cluster
Launching Spark Applications
Resource Scheduling
Executors Scheduling
Monitoring and Logging
Running Alongside Hadoop
Configuring Ports for Network Security
High Availability
Standby Masters with ZooKeeper
Single-Node Recovery with Local File System
In addition to running on the Mesos or YARN cluster managers, Spark also provides a simple standalone deploy mode. You can launch a standalone cluster either manually, by starting a master and workers by hand, or use our provided launch scripts. It is also possible to run these daemons on a single machine for testing.
Installing Spark Standalone to a Cluster
To install Spark Standalone mode, you simply place a compiled version of Spark on each node on the cluster. You can obtain pre-built versions of Spark with each release or build it yourself.
Starting a Cluster Manually
You can start a standalone master server by executing:
./sbin/start-master.sh
Once started, the master will print out a spark://HOST:PORT URL for itself, which you can use to connect workers to it, or pass as the "master"

Figure 6-2. Spark Standalone website

The previous command initializes the master node. You can access the master node interface at *localhost:8080*, as captured in Figure 6-3. Note that the Spark master URL is specified as *spark://address:port*; you will need this URL to initialize worker nodes.

We then can initialize a single worker using the master URL; however, you could use a similar approach to initialize multiple workers by running the code multiple times and, potentially, across different machines:

```
# Start worker node, find master URL at http://localhost:8080/
system2(spark_path, c("org.apache.spark.deploy.worker.Worker",
                      "spark://address:port"), wait = FALSE)
```

Figure 6-3. The Spark Standalone web interface

There is one worker register in Spark Standalone. Click the link to this worker node to view details for this particular worker, like available memory and cores, as shown in Figure 6-4.

Figure 6-4. Spark Standalone worker web interface

After you are done performing computations in this cluster, you will need to stop the master and worker nodes. You can use the `jps` command to identify the process numbers to terminate. In the following example, 15330 and 15353 are the processes that you can terminate to finalize this cluster. To terminate a process, you can use `system("Taskkill /PID ##### /F")` in Windows, or `system("kill -9 #####")` in macOS and Linux.

```
system("jps")
```

```
15330 Master
15365 Jps
15353 Worker
1689 QuorumPeerMain
```

You can follow a similar approach to configure a cluster by running the initialization code over each machine in the cluster.

While it's possible to initialize a simple standalone cluster, configuring a proper Spark Standalone cluster that can recover from computer restarts and failures, and supports multiple users, permissions, and so on, is usually a much longer process that falls beyond the scope of this book. The following sections present several alternatives that can be much easier to manage on-premises or through cloud services. We will start by introducing YARN.

YARN

Hadoop YARN, or simply YARN, as it is commonly called, is the resource manager of the Hadoop project. It was originally developed in the Hadoop project but was refactored into its own project in Hadoop 2. As we mentioned in Chapter 1, Spark was built to speed up computation over Hadoop, and therefore it's very common to find Spark installed on Hadoop clusters.

One advantage of YARN is that it is likely to be already installed in many existing clusters that support Hadoop; this means that you can easily use Spark with many existing Hadoop clusters without requesting any major changes to the existing cluster infrastructure. It is also very common to find Spark deployed in YARN clusters since many started out as Hadoop clusters and were eventually upgraded to also support Spark.

You can submit YARN applications in two modes: *yarn-client* and *yarn-cluster*. In yarn-cluster mode the driver is running remotely (potentially), while in yarn-client mode, the driver is running locally. Both modes are supported, and we explain them further in Chapter 7.

YARN provides a resource management user interface useful to access logs, monitor available resources, terminate applications, and more. After you connect to Spark from R, you will be able to manage the running application in YARN, as shown in Figure 6-5.

Since YARN is the cluster manager from the Hadoop project, you can find YARN's documentation at hadoop.apache.org (*http://bit.ly/2TDGsCX*). You can also reference the "Running Spark on YARN" guide at spark.apache.org (*http://bit.ly/306WsQx*).

Figure 6-5. YARN's Resource Manager running a sparklyr application

Apache Mesos

Apache Mesos is an open source project to manage computer clusters. Mesos began as a research project in the UC Berkeley RAD Lab. It makes use of Linux Cgroups (*http://bit.ly/2Z9KEeW*) to provide isolation for CPU, memory, I/O, and file system access.

Mesos, like YARN, supports executing many cluster frameworks, including Spark. However, one advantage particular to Mesos is that it allows cluster frameworks like Spark to implement custom task schedulers. A scheduler is the component that coordinates in a cluster which applications are allocated execution time and which resources are assigned to them. Spark uses a coarse-grained scheduler (*https://oreil.ly/9WQvg*), which schedules resources for the duration of the application; however, other frameworks might use Mesos' fine-grained scheduler, which can increase the overall efficiency in the cluster by scheduling tasks in shorter intervals, allowing them to share resources between them.

Mesos provides a web interface to manage your running applications, resources, and so on. After connecting to Spark from R, your application will be registered like any other application running in Mesos. Figure 6-6 shows a successful connection to Spark from R.

Mesos is an Apache project with its documentation available at mesos.apache.org (*https://mesos.apache.org/*). The *Running Spark on Mesos* (*http://bit.ly/31H4LCT*) guide is also a great resource if you choose to use Mesos as your cluster manager.

Figure 6-6. Mesos web interface running Spark and R

Distributions

You can use a cluster manager in on-premises clusters, as described in the previous section; however, many organizations—including, but not limited to, Apache Spark— choose to partner with companies providing additional management software, services, and resources to help manage applications in their cluster. Some of the on-premises cluster providers include *Cloudera*, *Hortonworks*, and *MapR*, which we briefly introduce next.

Cloudera, Inc., is a US-based software company that provides Apache Hadoop and Apache Spark–based software, support and services, and training to business customers. Cloudera's hybrid open source Apache Hadoop distribution, Cloudera Distribution Including Apache Hadoop (CDH), targets enterprise-class deployments of that technology. Cloudera donates more than 50% of its engineering output to the various Apache-licensed open source projects (Apache Hive, Apache Avro, Apache HBase, and so on) that combine to form the Apache Hadoop platform. Cloudera (*http://bit.ly/2KJmcfe*) is also a sponsor of the Apache Software Foundation.

Cloudera clusters make use of *parcels* (*http://bit.ly/33LHpxU*), which are binary distributions containing program files and metadata. Spark happens to be installed as a parcel in Cloudera. It's beyond the scope of this book to present how to configure Cloudera clusters, but resources and documentation can be found under cloudera.com (*http://bit.ly/33yUUkp*), and "Introducing sparklyr, an R Interface for Apache Spark" (*http://bit.ly/2HbAtjY*) on the Cloudera blog.

Cloudera provides the Cloudera Manager web interface to manage resources, services, parcels, diagnostics, and more. Figure 6-7 shows a Spark parcel running in Cloudera Manager, which you can later use to connect from R.

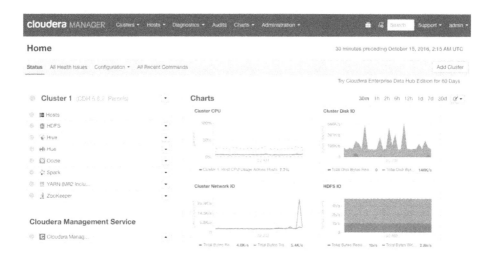

Figure 6-7. Cloudera Manager running Spark parcel

sparklyr is certified with Cloudera (*http://bit.ly/2z1yydc*), meaning that Cloudera's support is aware of sparklyr and can be effective helping organizations that are using Spark and R. Table 6-1 summarizes the versions currently certified.

Table 6-1. Versions of sparklyr certified with Cloudera

Cloudera version	Product	Version	Components	Kerberos
CDH5.9	sparklyr	0.5	HDFS, Spark	Yes
CDH5.9	sparklyr	0.6	HDFS, Spark	Yes
CDH5.9	sparklyr	0.7	HDFS, Spark	Yes

Hortonworks is a big data software company based in Santa Clara, California. The company develops, supports, and provides expertise on an expansive set of entirely open source software designed to manage data and processing for everything from Internet of Things (IoT) to advanced analytics and machine learning. Hortonworks (*http://bit.ly/2KTufpV*) believes that it is a data management company bridging the cloud and the datacenter.

Hortonworks partnered with Microsoft (*http://bit.ly/2NbfuBH*) to improve support in Microsoft Windows for Hadoop and Spark, which used to be a differentiation point from Cloudera; however, comparing Hortonworks and Cloudera is less relevant today since the companies merged in January 2019 (*http://bit.ly/2Mk1UMt*). Despite the merger, support for the Cloudera and Hortonworks Spark distributions are still available. Additional resources to configure Spark under Hortonworks are available at hortonworks.com (*http://bit.ly/2Z8M8Kh*).

MapR is a business software company headquartered in Santa Clara, California. MapR (*http://bit.ly/33DU8Cs*) provides access to a variety of data sources from a single computer cluster, including big data workloads such as Apache Hadoop and Apache Spark, a distributed file system, a multimodel database management system, and event stream processing, combining analytics in real time with operational applications. Its technology runs on both commodity hardware and public cloud computing services.

Cloud

If you have neither an on-premises cluster nor spare machines to reuse, starting with a cloud cluster can be quite convenient since it will allow you to access a proper cluster in a matter of minutes. This section briefly mentions some of the major cloud infrastructure providers and gives you resources to help you get started if you choose to use a cloud provider.

In cloud services, the compute instances are billed for as long the Spark cluster runs; your billing starts when the cluster launches, and it stops when the cluster stops. This cost needs to be multiplied by the number of instances reserved for your cluster. So, for instance, if a cloud provider charges $1.00 per compute instance per hour, and you start a three-node cluster that you use for one hour and 10 minutes, it is likely that you'll receive a bill for $1.00 x 2 hours x 3 nodes = $6.00. Some cloud providers charge per minute, but at least you can rely on all of them charging per compute hour.

Be aware that, while computing costs can be quite low for small clusters, accidentally leaving a cluster running can cause significant billing expenses. Therefore, it's worth taking the extra time to check twice that your cluster is terminated when you no longer need it. It's also a good practice to monitor costs daily while using clusters to make sure your expectations match the daily bill.

From past experience, you should also plan to request compute resources in advance while dealing with large-scale projects; various cloud providers will not allow you to start a cluster with hundreds of machines before requesting them explicitly through a support request. While this can be cumbersome, it's also a way to help you control costs in your organization.

Since the cluster size is flexible, it is a good practice to start with small clusters and scale compute resources as needed. Even if you know in advance that a cluster of significant size will be required, starting small provides an opportunity to troubleshoot issues at a lower cost since it's unlikely that your data analysis will run at scale flawlessly on the first try. As a rule of thumb, grow the instances exponentially; if you need to run a computation over an eight-node cluster, start with one node and an eighth of the entire dataset, then two nodes with a fourth, then four nodes with half of the dataset, and then, finally, eight nodes and the entire dataset. As you become more experienced, you'll develop a good sense of how to troubleshoot issues, and of the size of the required cluster, and you'll be able to skip intermediate steps, but for starters, this is a good practice to follow.

You can also use a cloud provider to acquire bare computing resources and then install the on-premises distributions presented in the previous section yourself; for instance, you can run the Cloudera distribution on Amazon Elastic Compute Cloud (Amazon EC2). This model would avoid procuring colocated hardware, but it still allows you to closely manage and customize your cluster. This book presents an overview of only the fully managed Spark services available by cloud providers; however, you can usually find with ease instructions online on how to install on-premises distributions in the cloud.

Some of the major providers of cloud computing infrastructure are Amazon, Databricks, Google, IBM, and Microsoft and Qubole. The subsections that follow briefly introduce each one.

Amazon

Amazon provides cloud services through AWS (*https://aws.amazon.com/*); more specifically, it provides an on-demand Spark cluster through Amazon EMR (*https://aws.amazon.com/emr/*).

Detailed instructions on using R with Amazon EMR were published under Amazon's Big Data blog in a post called "Running sparklyr on Amazon EMR" (*https://amzn.to/2OYWMQ5*). This post introduced the launch of sparklyr and instructions to configure Amazon EMR clusters with sparklyr. For instance, it suggests you can use the Amazon Command Line Interface (*https://aws.amazon.com/cli/*) to launch a cluster with three nodes, as follows:

```
aws emr create-cluster --applications Name=Hadoop Name=Spark Name=Hive \
  --release-label emr-5.8.0 --service-role EMR_DefaultRole --instance-groups \
  InstanceGroupType=MASTER,InstanceCount=1,InstanceType=m3.2xlarge \
  InstanceGroupType=CORE,InstanceCount=2,InstanceType=m3.2xlarge \
  --bootstrap-action Path=s3://aws-bigdata-blog/artifacts/aws-blog-emr-\
rstudio-sparklyr/rstudio_sparklyr_emr5.sh,Args=["--user-pw", "<password>", \
  "--rstudio", "--arrow"] --ec2-attributes InstanceProfile=EMR_EC2_DefaultRole
```

You can then see the cluster launching and then eventually running under the AWS portal, as illustrated in Figure 6-8.

You then can navigate to the Master Public DNS and find RStudio under port 8787—for example, ec2-12-34-567-890.us-west-1.compute.amazonaws.com:8787—and then log in with user hadoop and password password.

Figure 6-8. Launching an Amazon EMR cluster

It is also possible to launch the Amazon EMR cluster using the web interface; the same introductory post contains additional details and walkthroughs specifically designed for Amazon EMR.

Remember to turn off your cluster to avoid unnecessary charges and use appropriate security restrictions when starting Amazon EMR clusters for sensitive data analysis.

Regarding cost, you can find the most up-to-date information at Amazon EMR Pricing (*https://amzn.to/2YRGb5r*). Table 6-2 presents some of the instance types available in the us-west-1 region (as of this writing); this is meant to provide a glimpse of the resources and costs associated with cloud processing. Notice that the "EMR price is in addition to the Amazon EC2 price (the price for the underlying servers)."

Table 6-2. Amazon EMR pricing information

Instance	CPUs	Memory	Storage	EC2 cost	EMR cost
c1.medium	2	1.7 GB	350 GB	$0.148/hour	$0.030/hour
m3.2xlarge	8	30 GB	160 GB	$0.616/hour	$0.140/hour
i2.8xlarge	32	244 GB	6400 GB	$7.502/hour	$0.270/hour

We are presenting only a subset of the available compute instances for Amazon and subsequent cloud providers as of 2019; however, note that hardware (CPU speed, hard drive speed, etc.) varies between vendors and locations; therefore, you can't use these hardware tables as an accurate price comparison. Accurate comparison would require running your particular workloads and considering other aspects beyond compute instance cost.

Databricks

Databricks (*https://databricks.com*) is a company founded by the creators of Apache Spark, whose aim is to help clients with cloud-based big data processing using Spark. Databricks grew out of the AMPLab (*https://oreil.ly/W2Eoe*) project at the University of California, Berkeley.

Databricks provides enterprise-level cluster computing plans as well as a free/community tier to explore functionality and become familiar with their environment.

After a cluster is launched, you can use R and `sparklyr` from Databricks notebooks following the steps provided in Chapter 2 or by installing RStudio on Databricks (*http://bit.ly/2KCDax6*). Figure 6-9 shows a Databricks notebook using Spark through `sparkylyr`.

Figure 6-9. Databricks community notebook running sparklyr

Additional resources are available under the Databricks Engineering Blog post "Using sparklyr in Databricks" (*http://bit.ly/2N59jyR*) and the Databricks documentation for `sparklyr` (*http://bit.ly/2MkOYWC*).

You can find the latest pricing information at *databricks.com/product/pricing* (*http://bit.ly/305Rnrt*). Table 6-3 lists the available plans as of this writing.

Table 6-3. Databricks procong information

Plan	Basic	Data engineering	Data analytics
AWS Standard	$0.07/DBU	$0.20/DBU	$0.40/DBU
Azure Standard		$0.20/DBU	$0.40/DBU
Azure Premium		$0.35/DBU	$0.55/DBU

Notice that pricing is based on cost of DBU per hour. From Databricks, "a Databricks Unit (*https://oreil.ly/3muQq*) (DBU) is a unit of Apache Spark processing capability per hour. For a varied set of instances, DBUs are a more transparent way to view usage instead of the node-hour."

Google

Google provides Google Cloud Dataproc as a cloud-based managed Spark and Hadoop service offered on Google Cloud Platform (GCP). Dataproc utilizes many GCP technologies, such as Google Compute Engine and Google Cloud Storage, to offer fully managed clusters running popular data processing frameworks such as Apache Hadoop and Apache Spark.

You can easily create a cluster from the Google Cloud console or the Google Cloud command-line interface (CLI) as illustrated in Figure 6-10.

Figure 6-10. Launching a Dataproc cluster

After you've created your cluster, ports can be forwarded to allow you to access this cluster from your machine—for instance, by launching Chrome to make use of this proxy and securely connect to the Dataproc cluster. Configuring this connection looks as follows:

```
gcloud compute ssh sparklyr-m --project=<project> --zone=<region> -- -D 1080 \
  -N "<path to chrome>" --proxy-server="socks5://localhost:1080" \
  --user-data-dir="/tmp/sparklyr-m" http://sparklyr-m:8088
```

There are various tutorials available (*http://bit.ly/2OYyo18*) (cloud.google.com/dataproc/docs/tutorials), including a comprehensive tutorial to configure RStudio and sparklyr (*http://bit.ly/2MhSgKg*).

You can find the latest pricing information at *cloud.google.com/dataproc/pricing* (*http://bit.ly/31J0uyC*). In Table 6-4 notice that the cost is split between compute engine and a dataproc premium.

Table 6-4. Google Cloud Dataproc pricing information

Instance	CPUs	Memory	Compute engine	Dataproc premium
n1-standard-1	1	3.75 GB	$0.0475/hour	$0.010/hour
n1-standard-8	8	30 GB	$0.3800/hour	$0.080/hour
n1-standard-64	64	244 GB	$3.0400/hour	$0.640/hour

IBM

IBM cloud computing is a set of cloud computing services for business. IBM cloud includes Infrastructure as a Service (IaaS), Software as a Service (SaaS), and Platform as a Service (PaaS) offered through public, private, and hybrid cloud delivery models, in addition to the components that make up those clouds.

From within IBM Cloud, open Watson Studio and create a Data Science project, add a Spark cluster under the project settings, and then, on the Launch IDE menu, start RStudio. Please note that, as of this writing, the provided version of sparklyr was not the latest version available in CRAN, since sparklyr was modified to run under the IBM Cloud. In any case, follow IBM's documentation as an authoritative reference to run R and Spark on the IBM Cloud and particularly on how to upgrade sparklyr appropriately. Figure 6-11 captures IBM's Cloud portal launching a Spark cluster.

Figure 6-11. IBM Watson Studio launching Spark with R support

The most up-to-date pricing information is available at *ibm.com/cloud/pricing* (*https://www.ibm.com/cloud/pricing*). In Table 6-5, compute cost was normalized using 31 days from the per-month costs.

Table 6-5. IBM Cloud pricing information

Instance	CPUs	Memory	Storage	Cost
C1.1x1x25	1	1 GB	25 GB	$0.033/hour
C1.4x4x25	4	4 GB	25 GB	$0.133/hour
C1.32x32x25	32	25 GB	25 GB	$0.962/hour

Microsoft

Microsoft Azure is a cloud computing service created by Microsoft for building, testing, deploying, and managing applications and services through a global network of Microsoft-managed datacenters. It provides SaaS, PaaS, and IaaS and supports many different programming languages, tools, and frameworks, including both Microsoft-specific and third-party software and systems.

From the Azure portal, the Azure HDInsight service provides support for on-demand Spark clusters. You can easily create HDInsight cluster with support for Spark and RStudio by selecting the ML Services cluster type. Note that the provided version of `sparklyr` might not be the latest version available in CRAN since the default package repository seems to be initialized using a Microsoft R Application

Network (MRAN) snapshot, not directly from CRAN. Figure 6-12 shows the Azure portal launching a Spark cluster with support for R.

Figure 6-12. Creating an Azure HDInsight Spark cluster

Up-to-date pricing for HDInsight is available at *azure.microsoft.com/en-us/pricing/details/hdinsight* (*http://bit.ly/2H9Ce0X*); Table 6-6 lists the pricing as of this writing.

Table 6-6. Azure HDInsight pricing information

Instance	CPUs	Memory	Total cost
D1 v2	1	3.5 GB	$0.074/hour
D4 v2	8	28 GB	$0.59/hour
G5	64	448 GB	$9.298/hour

Qubole

Qubole (*https://www.qubole.com*) was founded in 2013 with a mission to close the data accessibility gap. Qubole delivers a self-service platform for big data analytics built on Amazon, Microsoft, Google, and Oracle Clouds. In Qubole, you can launch Spark clusters, which you can then use from Qubole notebooks (*http://bit.ly/33ChKYk*) or RStudio Server. Figure 6-13 shows a Qubole cluster initialized with RStudio and `sparklyr`.

Figure 6-13. A Qubole cluster initialized with RStudio and sparklyr

You can find the latest pricing information at Qubole's pricing page (*http://bit.ly/33AuKh8*). Table 6-7 lists the price for Qubole's current plan, as of this writing. Notice that pricing is based on cost of QCU/hr, which stands for "Qubole Compute Unit per hour," and the Enterprise Edition requires an annual contract.

Table 6-7. Qubole pricing information

Test Drive	Full-featured trial	Enterprise Edition
$0	$0	$0.14/QCU

Kubernetes

Kubernetes is an open source container orchestration system for automating deployment, scaling, and management of containerized applications that was originally designed by Google and is now maintained by the Cloud Native Computing Foundation (*https://www.cncf.io/*) (CNCF). Kubernetes was originally based on Docker (*https://www.docker.com/*), while, like Mesos, it's also based on Linux Cgroups.

Kubernetes can execute many cluster applications and frameworks that you can highly customize by using container images with specific resources and libraries. This allows a single Kubernetes cluster to be used for many different purposes beyond data analysis, which in turn helps organizations manage their compute resources with ease. One trade-off from using custom images is that they add further configuration overhead but make Kubernetes clusters extremely flexible. Nevertheless, this flexibility has proven to be instrumental to easily administer cluster resources in many organizations and, as pointed out in "Overview" on page 93, Kubernetes is becoming a very popular cluster framework.

Kubernetes is supported across all major cloud providers. They all provide extensive documentation as to how to launch, manage, and tear down Kubernetes clusters; Figure 6-14 shows the GCP console while creating a Kubernetes cluster. You can deploy Spark over any Kubernetes cluster, and you can use R to connect, analyze, model, and more.

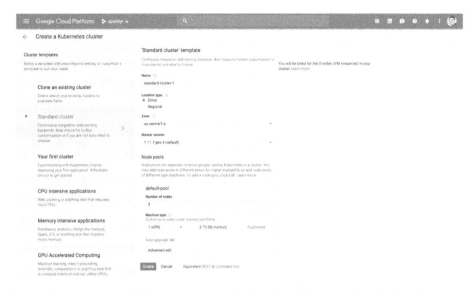

Figure 6-14. Creating a Kubernetes cluster for Spark and R using Google Cloud

You can learn more at kubernetes.io (*https://kubernetes.io/*), and read the *Running Spark on Kubernetes* guide from spark.apache.org (*http://bit.ly/2KAZze7*).

Strictly speaking, Kubernetes is a cluster technology, not a specific cluster architecture. However, Kubernetes represents a larger trend often referred to as a *hybrid cloud*. A hybrid cloud is a computing environment that makes use of on-premises and public cloud services with orchestration between the various platforms. It's still too early to precisely categorize the leading technologies that will form a hybrid approach to cluster computing; although, as previously mentioned, Kubernetes is the leading one, many more are likely to form to complement or even replace existing technologies.

Tools

While using only R and Spark can be sufficient for some clusters, it is common to install complementary tools in your cluster to improve monitoring, SQL analysis, workflow coordination, and more, with applications like Ganglia (*http://ganglia.info/*), Hue (*http://gethue.com/*), and Oozie (*https://oozie.apache.org*), respectively. This section is not meant to cover all tools; rather, it mentions the ones that are commonly used.

RStudio

From reading Chapter 1, you are aware that RStudio is a well-known and free desktop development environment for R; therefore, it is likely that you are following the examples in this book using RStudio Desktop. However, you might not be aware that you can run RStudio as a web service within a Spark cluster. This version of RStudio is known as *RStudio Server*. You can see RStudio Server running in Figure 6-15. In the same way that the Spark UI runs in the cluster, you can install RStudio Server within the cluster. Then you can connect to RStudio Server and use RStudio in exactly the same way you use RStudio Desktop but with the ability to run code against the Spark cluster. As you can see in Figure 6-15, RStudio Server looks and feels just like RStudio Desktop, but adds support to run commands efficiently by being located within the cluster.

Figure 6-15. RStudio Server Pro running in AWS

If you're familiar with R, Shiny Server is a very popular tool for building interactive web applications from R. We recommended that you install Shiny directly in your Spark cluster.

RStudio Server and Shiny Server are a free and open source; however, RStudio also provides professional products like RStudio Server, RStudio Server Pro (*http://bit.ly/2KCaxQn*), Shiny Server Pro (*http://bit.ly/30aV0fK*), and RStudio Connect (*http://bit.ly/306fHcY*), which you can install within the cluster to support additional R workflows. While sparklyr does not require any additional tools, they provide significant productivity gains worth considering. You can learn more about them at *rstudio.com/products/* (*http://bit.ly/2MihHLP*).

Jupyter

Project Jupyter (*http://jupyter.org/*) exists to develop open source software, open standards, and services for interactive computing across dozens of programming languages. A Jupyter notebook provides support for various programming languages, including R. You can use `sparklyr` with Jupyter notebooks using the R Kernel. Figure 6-16 shows `sparklyr` running within a local Jupyter notebook.

Figure 6-16. Jupyter notebook running sparklyr

Livy

Apache Livy (*http://bit.ly/2L2TZAn*) is an incubation project in Apache providing support to use Spark clusters remotely through a web interface. It is ideal to connect directly into the Spark cluster; however, there are times where connecting directly to the cluster is not feasible. When facing those constraints, you can consider installing Livy in the cluster and secure it properly to enable remote use over web protocols. Be aware, though, that there is a significant performance overhead from using Livy in `sparklyr`.

To help test Livy locally, `sparklyr` provides support to list, install, start, and stop a local Livy instance by executing `livy_available_versions()`:

```
##     livy
## 1 0.2.0
## 2 0.3.0
## 3 0.4.0
## 4 0.5.0
```

This lists the versions that you can install; we recommend installing the latest version and verifying it as follows:

```
# Install default Livy version
livy_install()

# List installed Livy services
livy_installed_versions()

# Start the Livy service
livy_service_start()
```

You then can navigate to the local Livy session at *http://localhost:8998*. Chapter 7 will detail how to connect through Livy. After you're connected, you can navigate to the Livy web application, as shown in Figure 6-17.

Figure 6-17. Apache Livy running as a local service

Make sure you also stop the Livy service when working with local Livy instances (for proper Livy services running in a cluster, you won't have to):

```
# Stops the Livy service
livy_service_stop()
```

Recap

This chapter explained the history and trade-offs of on-premises and cloud computing and presented Kubernetes as a promising framework to provide flexibility across on-premises or multiple cloud providers. It also introduced cluster managers (Spark Standalone, YARN, Mesos, and Kubernetes) as the software needed to run Spark as a cluster application. This chapter briefly mentioned on-premises cluster providers like Cloudera, Hortonworks, and MapR, as well as the major Spark cloud providers: Amazon, Databricks, IBM, Google, Microsoft, and Qubole.

While this chapter provided a solid foundation to understand current cluster computing trends, tools, and service providers useful to perform data science at scale, it did not provide a comprehensive framework to help you decide which cluster technologies to choose. Instead, use this chapter as an overview and a starting point to seek out additional resources to help you find the cluster stack that best fits your organization needs.

Chapter 7 will focus on understanding how to connect to existing clusters; therefore, it assumes a Spark cluster like those we presented in this chapter is already available to you.

Connections

They don't get to choose.

—Daenerys Targaryen

Chapter 6 presented the major cluster computing trends, cluster managers, distributions, and cloud service providers to help you choose the Spark cluster that best suits your needs. In contrast, this chapter presents the internal components of a Spark cluster and how to connect to a particular Spark cluster.

When reading this chapter, don't try to execute every line of code; this would be quite hard since you would need to prepare different Spark environments. Instead, if you already have a Spark cluster or if the previous chapter gets you motivated enough to sign up for an on-demand cluster, now is the time to learn how to connect to it. This chapter helps you connect to your cluster, which you should have already chosen by now. Without a cluster, we recommend that you learn the concepts and come back to execute code later on.

In addition, this chapter provides various troubleshooting connection techniques. While we hope you won't need to use them, this chapter prepares you to use them as effective techniques to resolve connectivity issues.

While this chapter might feel a bit dry—connecting and troubleshooting connections is definitely not the most exciting part of large-scale computing—it introduces the components of a Spark cluster and how they interact, often known as the *architecture* of Apache Spark. This chapter, along with Chapters 8 and 9, will provide a detailed view of how Spark works, which will help you move toward becoming an intermediate Spark user who can truly understand the exciting world of distributed computing using Apache Spark.

Overview

The overall connection architecture for a Spark cluster is composed of three types of compute instances: the *driver node*, the *worker nodes*, and the *cluster manager*. A cluster manager is a service that allows Spark to be executed in the cluster; this was detailed in "Managers" on page 96. The worker nodes (also referred to as *executors*) execute compute tasks over partitioned data and communicate intermediate results to other workers or back to the driver node. The driver node is tasked with delegating work to the worker nodes, but also with aggregating their results and controlling computation flow. For the most part, aggregation happens in the worker nodes; however, even after the nodes aggregate data, it is often the case that the driver node would need to collect the worker's results. Therefore, the driver node usually has at least, but often much more, compute resources (memory, CPUs, local storage, etc.) as the worker node.

Strictly speaking, the driver node and worker nodes are just names assigned to machines with particular roles, while the actual computation in the driver node is performed by the *Spark context*. The Spark context is the main entry point for Spark functionality since it's tasked with scheduling tasks, managing storage, tracking execution status, specifying access configuration settings, canceling jobs, and so on. In the worker nodes, the actual computation is performed under a *spark executor*, which is a Spark component tasked with executing subtasks against a specific data partition.

Figure 7-1 illustrates this concept, where the driver node orchestrates a worker's work through the cluster manager.

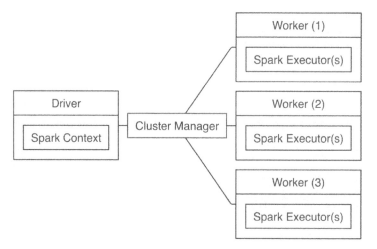

Figure 7-1. Apache Spark connection architecture

If you already have a Spark cluster in your organization, you should request the connection information to this cluster from your cluster administrator, read their usage policies carefully, and follow their advice. Since a cluster can be shared among many users, you want to ensure that you request only the compute resources you need. We cover how to request resources in Chapter 9. Your system administrator will specify whether it's an *on-premises* or *cloud* cluster, the cluster manager being used, supported *connections*, and supported *tools*. You can use this information to jump directly to Local, Standalone, YARN, Mesos, Livy, or Kubernetes based on which is appropriate for your situation.

 After you've used spark_connect() to connect, you can use all the techniques described in previous chapters using the sc connection; for instance, you can do data analysis or modeling with the same code previous chapters presented.

Edge Nodes

Computing clusters are configured to enable high bandwidth and fast network connectivity between nodes. To optimize network connectivity, the nodes in the cluster are configured to trust one another and to disable security features. This improves performance but requires you to close all external network communication, making the entire cluster secure as a whole except for a few cluster machines that are carefully configured to accept connections from outside the cluster; conceptually, these machines are located in the "edge" of the cluster and are known as *edge nodes*.

Therefore, before connecting to Apache Spark, it is likely that you will first need to connect to an edge node in your cluster. There are two methods to connect:

Terminal
Using a computer terminal application, you can use a Secure Shell (*http://bit.ly/ 2TE8cY9*) to establish a remote connection into the cluster; after you connect into the cluster, you can launch R and then use sparklyr. However, a terminal can be cumbersome for some tasks, like exploratory data analysis, so it's often used only while configuring the cluster or troubleshooting issues.

Web browser
While using sparklyr from a terminal is possible, it is usually more productive to install a *web server* in an edge node that provides access to run R with spar klyr from a web browser. Most likely, you will want to consider using RStudio or Jupyter rather than connecting from the terminal.

Figure 7-2 explains these concepts visually. The left block is usually your web browser, and the right block is the edge node. Client and edge nodes communicate over HTTP when using a web browser or Secure Shell (SSH) when using the terminal.

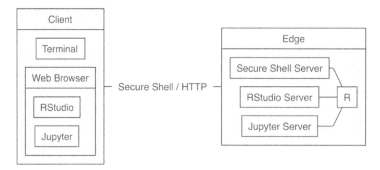

Figure 7-2. Connecting to Spark's edge node

Spark Home

After you connect to an edge node, the next step is to determine where Spark is installed, a location known as the SPARK_HOME. In most cases, your cluster administrator will have already set the SPARK_HOME environment variable to the correct installation path. If not, you will need to get the correct *SPARK_HOME* path. You must specify the SPARK_HOME path as an environment variable or explicitly when running spark_connect() using the spark_home parameter.

If your cluster provider or cluster administrator already provided SPARK_HOME for you, the following code should return a path instead of an empty string:

```
Sys.getenv("SPARK_HOME")
```

If this code returns an empty string, this would mean that the SPARK_HOME environment variable is not set in your cluster, so you will need to specify SPARK_HOME while using spark_connect(), as follows:

```
sc <- spark_connect(master = "master", spark_home = "local/path/to/spark")
```

In this example, master is set to the correct cluster manager master for Spark Standalone, YARN, Mesos, Kubernetes, or Livy.

Local

When you connect to Spark in local mode, Spark starts a single process that runs most of the cluster components like the Spark context and a single executor. This is ideal to learn Spark, work offline, troubleshoot issues, or test code before you run it over a large compute cluster. Figure 7-3 depicts a local connection to Spark.

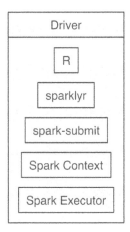

Figure 7-3. Local connection diagram

Notice that there is neither a cluster manager nor worker process since, in local mode, everything runs inside the driver application. It's also worth noting that spar klyr starts the Spark context through spark-submit, a script available in every Spark installation to enable users to submit custom applications to Spark. If you're curious, Chapter 13 explains the internal processes that take place in sparklyr to submit this application and connect properly from R.

To perform this local connection, we can use the following familiar code from previous chapters:

```
# Connect to local Spark instance
sc <- spark_connect(master = "local")
```

Standalone

Connecting to a Spark Standalone cluster requires the location of the cluster manager's master instance, which you can find in the cluster manager web interface as described in the "Standalone" on page 96 section. You can find this location by looking for a URL starting with spark://.

A connection in standalone mode starts from sparklyr, which launches spark-submit, which then submits the sparklyr application and creates the Spark Context, which requests executors from the Spark Standalone instance running under the given master address.

Figure 7-4 illustrates this process, which is quite similar to the overall connection architecture from Figure 7-1 but with additional details that are particular to standalone clusters and sparklyr.

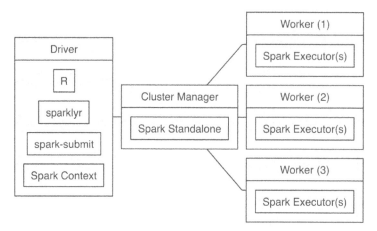

Figure 7-4. Spark Standalone connection diagram

To connect, use `master = "spark://hostname:port"` in `spark_connect()` as follows:

```
sc <- spark_connect(master = "spark://hostname:port")
```

YARN

Hadoop YARN is the cluster manager from the Hadoop project. It's the most common cluster manager that you are likely to find in clusters, which started out as Hadoop clusters; with Cloudera, Hortonworks, and MapR distributions as when using Amazon EMR. YARN supports two connection modes: YARN client and YARN cluster. However, YARN client mode is much more common than YARN cluster since it's more efficient and easier to set up.

YARN Client

When you connect in YARN client mode, the driver instance runs R, `sparklyr`, and the Spark context, which requests worker nodes from YARN to run Spark executors, as shown in Figure 7-5.

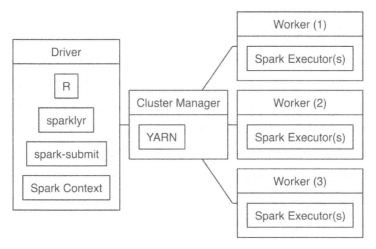

Figure 7-5. YARN client connection diagram

To connect, you simply run with `master = "yarn"`, as follows:

```
sc <- spark_connect(master = "yarn")
```

Behind the scenes, when you're running YARN in client mode, the cluster manager will do what you would expect a cluster manager would do: it allocates resources from the cluster and assigns them to your Spark application, which the Spark context will manage for you. The important piece to notice in "YARN" on page 122 is that the Spark context resides in the same machine where you run R code; this is different when you're running YARN in cluster mode.

YARN Cluster

The main difference between running YARN in cluster mode and running YARN in client mode is that, in cluster mode, the driver node is not required to be the node where R and `sparklyr` were launched; instead, the driver node remains the designated driver node, which is usually a different node than the edge node where R is running. It can be helpful to consider using cluster mode when the edge node has too many concurrent users, when it is lacking computing resources, or when tools (like RStudio or Jupyter) need to be managed independently of other cluster resources.

Figure 7-6 shows how the different components become decoupled when running in cluster mode. Notice there is still a line connecting the client with the cluster manager since, first of all, resources still need to be allocated from the cluster manager; however, after they're allocated, the client communicates directly with the driver node, which communicates with the worker nodes. From Figure 7-6, you might think that cluster mode looks much more complicated than client mode—this would be a correct assessment; therefore, if possible, it's best to avoid cluster mode due to its additional configuration overhead.

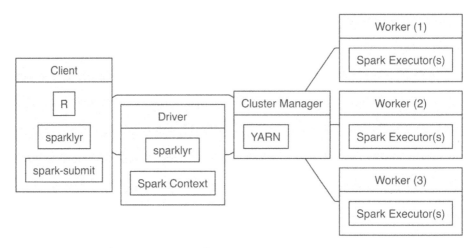

Figure 7-6. YARN cluster connection diagram

To connect in YARN cluster mode, simply run the following:

```
sc <- spark_connect(master = "yarn-cluster")
```

Cluster mode assumes that the node running `spark_connect()` is properly configured, meaning that `yarn-site.xml` exists and the `YARN_CONF_DIR` environment variable is properly set. When using Hadoop as a file system, you will also need the `HADOOP_CONF_DIR` environment variable properly configured. In addition, you would need to ensure proper network connectivity between the client and the driver node— not just by having both machines reachable, but also by making sure that they have sufficient bandwidth between them. This configuration is usually provided by your system administrator and is not something that you would need to manually configure.

Livy

As opposed to other connection methods that require using an edge node in the cluster, Livy provides a *web API* that makes the Spark cluster accessible from outside the cluster and does not require a Spark installation in the client. After it's connected through the web API, the *Livy Service* starts the Spark context by requesting resources from the cluster manager and distributing work as usual. Figure 7-7 illustrates a Livy connection; notice that the client connects remotely to the driver through a web API.

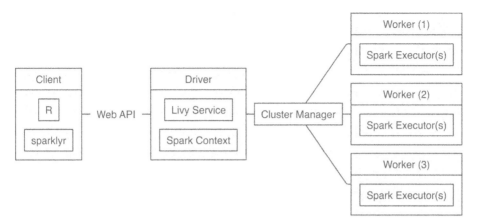

Figure 7-7. Livy connection diagram

Connecting through Livy requires the URL to the Livy service, which should be similar to `https://hostname:port/livy`. Since remote connections are allowed, connections usually require, at the very least, basic authentication:

```
sc <- spark_connect(
  master = "https://hostname:port/livy",
  method = "livy", config = livy_config(
    spark_version = "2.3.1",
    username = "<username>",
    password = "<password>"
  ))
```

To try out Livy on your local machine, you can install and run a Livy service as described under the "Livy" on page 114 section and then connect as follows:

```
sc <- spark_connect(
  master = "http://localhost:8998",
  method = "livy",
  version = "2.3.1")
```

After you're connected through Livy, you can make use of any `sparklyr` feature; however, Livy is not suitable for exploratory data analysis, since executing commands has a significant performance cost. That said, while running long-running computations, this overhead could be considered irrelevant. In general, you should prefer to avoid using Livy and work directly within an edge node in the cluster; when this is not feasible, using Livy could be a reasonable approach.

Specifying the Spark version through the `spark_version` parameter is optional; however, when the version is specified, performance is significantly improved by deploying precompiled Java binaries compatible with the given version. Therefore, it is a best practice to specify the Spark version when connecting to Spark using Livy.

Mesos

Similar to YARN, Mesos supports client mode and a cluster mode; however, `sparklyr` currently supports only client mode under Mesos. Therefore, the diagram shown in Figure 7-8 is equivalent to YARN client's diagram with only the cluster manager changed from YARN to Mesos.

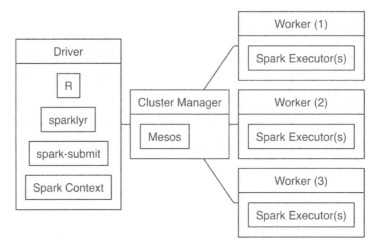

Figure 7-8. Mesos connection diagram

Connecting requires the address to the Mesos master node, usually in the form of `mesos://host:port` or `mesos://zk://host1:2181,host2:2181,host3:2181/mesos` for Mesos using ZooKeeper:

```
sc <- spark_connect(master = "mesos://host:port")
```

The `MESOS_NATIVE_JAVA_LIBRARY` environment variable needs to be set by your system administrator or manually set when you are running Mesos on your local machine. For instance, in macOS, you can install and initialize Mesos from a terminal, followed by manually setting the `mesos` library and connecting with `spark_connect()`:

```
brew install mesos
/usr/local/Cellar/mesos/1.6.1/sbin/mesos-master --registry=in_memory
    --ip=127.0.0.1 MESOS_WORK_DIR=. /usr/local/Cellar/mesos/1.6.1/sbin/mesos-slave
    --master=127.0.0.1:5050
```

```
Sys.setenv(MESOS_NATIVE_JAVA_LIBRARY =
           "/usr/local/Cellar/mesos/1.6.1/lib/libmesos.dylib")

sc <- spark_connect(master = "mesos://localhost:5050",
                    spark_home = spark_home_dir())
```

Kubernetes

Kubernetes clusters do not support client modes like Mesos or YARN; instead, the connection model is similar to YARN cluster, where the driver node is assigned by Kubernetes, as illustrated in Figure 7-9.

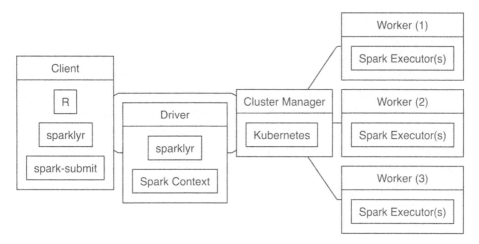

Figure 7-9. Kubernetes connection diagram

To use Kubernetes, you need to prepare a virtual machine with Spark installed and properly configured; however, it is beyond the scope of this book to present how to create one. Once created, connecting to Kubernetes works as follows:

```
library(sparklyr)
sc <- spark_connect(config = spark_config_kubernetes(
  "k8s://https://<apiserver-host>:<apiserver-port>",
  account = "default",
  image = "docker.io/owner/repo:version",
  version = "2.3.1"))
```

If your computer is already configured to use a Kubernetes cluster, you can use the following command to find the apiserver-host and apiserver-port:

```
system2("kubectl", "cluster-info")
```

Cloud

When you are working with cloud providers, there are a few connection differences. For instance, connecting from Databricks requires the following connection method:

```
sc <- spark_connect(method = "databricks")
```

Since Amazon EMR makes use of YARN, you can connect using `master = "yarn"`:

```
sc <- spark_connect(master = "yarn")
```

Connecting to Spark when using IBM's Watson Studio requires you to retrieve a configuration object through a `load_spark_kernels()` function that IBM provides:

```
kernels <- load_spark_kernels()
sc <- spark_connect(config = kernels[2])
```

In Microsoft Azure HDInsights and when using ML Services (R Server), a Spark connection is initialized as follows:

```
library(RevoScaleR)
cc <- rxSparkConnect(reset = TRUE, interop = "sparklyr")
sc <- rxGetSparklyrConnection(cc)
```

Connecting from Qubole requires using the `qubole` connection method:

```
sc <- spark_connect(method = "qubole")
```

Refer to your cloud provider's documentation and support channels if you need help.

Batches

Most of the time, you use `sparklyr` interactively; that is, you explicitly connect with `spark_connect()` and then execute commands to analyze and model large-scale data. However, you can also automate processes by scheduling Spark jobs that use `spar klyr`. Spark does not provide tools to schedule data-processing tasks; instead, you would use other workflow management tools. This can be useful to transform data, prepare a model and score data overnight, or to make use of Spark by other systems.

As an example, you can create a file named `batch.R` with the following contents:

```
library(sparklyr)

sc <- spark_connect(master = "local")

sdf_len(sc, 10) %>% spark_write_csv("batch.csv")

spark_disconnect(sc)
```

You then can submit this application to Spark in batch mode using `spark_submit()`; the `master` parameter should be set appropriately:

```
spark_submit(master = "local", "batch.R")
```

You can also invoke `spark-submit` from the shell directly through the following:

```
/spark-home-path/spark-submit
  --class sparklyr.Shell '/spark-jars-path/sparklyr-2.3-2.11.jar'
  8880 12345 --batch /path/to/batch.R
```

The last parameters represent the port number 8880 and the session number 12345, which you can set to any unique numeric identifier. You can use the following R code to get the correct paths:

```
# Retrieve spark-home-path
spark_home_dir()

# Retrieve spark-jars-path
system.file("java", package = "sparklyr")
```

You can customize your script by passing additional command-line arguments to `spark-submit` and then read them back in R using `commandArgs()`.

Tools

When connecting to a Spark cluster using tools like Jupyter and RStudio, you can run the same connection parameters presented in this chapter. However, since many cloud providers make use of a web proxy to secure Spark's web interface, to use `spark_web()` or the RStudio Connections pane extension, you need to properly configure the `sparklyr.web.spark` setting, which you would then pass to `spark_con fig()` through the `config` parameter.

For instance, when using Amazon EMR, you can configure `sparklyr.web.spark` and `sparklyr.web.yarn` by dynamically retrieving the YARN application and building the EMR proxy URL:

```
domain <- "http://ec2-12-345-678-9.us-west-2.compute.amazonaws.com"
config <- spark_config()
config$sparklyr.web.spark <- ~paste0(
  domain, ":20888/proxy/", invoke(spark_context(sc), "applicationId"))
config$sparklyr.web.yarn <- paste0(domain, ":8088")

sc <- spark_connect(master = "yarn", config = config)
```

Multiple Connections

It is common to connect once, and only once, to Spark. However, you can also open multiple connections to Spark by connecting to different clusters or by specifying the `app_name` parameter. This can be helpful to compare Spark versions or validate your analysis before submitting to the cluster. The following example opens connections to Spark 1.6.3, 2.3.0 and Spark Standalone:

```
# Connect to local Spark 1.6.3
sc_16 <- spark_connect(master = "local", version = "1.6")

# Connect to local Spark 2.3.0
sc_23 <- spark_connect(master = "local", version = "2.3", appName = "Spark23")

# Connect to local Spark Standalone
sc_standalone <- spark_connect(master = "spark://host:port")
```

Finally, you can disconnect from each connection:

```
spark_disconnect(sc_1_6_3)
spark_disconnect(sc_2_3_0)
spark_disconnect(sc_standalone)
```

Alternatively, you can disconnect from all connections at once:

```
spark_disconnect_all()
```

Troubleshooting

Last but not least, we introduce the following troubleshooting techniques: *Logging*, *Spark Submit*, and *Windows*. When in doubt about where to begin, start with the Windows section when using Windows systems, followed by Logging and finally Spark Submit. These techniques are useful when running `spark_connect()` fails with an error message.

Logging

The first technique to troubleshoot connections is to print Spark logs directly to the console to help you spot additional error messages:

```
sc <- spark_connect(master = "local", log = "console")
```

In addition, you can enable verbose logging by setting the `sparklyr.verbose` option to TRUE when connecting:

```
sc <- spark_connect(master = "local", log = "console",
                    config = list(sparklyr.verbose = TRUE))
```

Spark Submit

You can diagnose whether a connection issue is specific to R or Spark in general by running an example job through `spark-submit` and validating that no errors are thrown:

```
# Find the spark directory using an environment variable
spark_home <- Sys.getenv("SPARK_HOME")

# Or by getting the local spark installation
spark_home <- sparklyr::spark_home_dir()
```

Then, execute the sample compute Pi example by replacing "local" with the correct master parameter that you are troubleshooting:

```
# Launching a sample application to compute Pi
system2(
  file.path(spark_home, "bin", "spark-submit"),
  c(
    "--master", "local",
    "--class", "org.apache.spark.examples.SparkPi",
    dir(file.path(spark_home, "examples", "jars"),
        pattern = "spark-examples", full.names = TRUE),
    100),
  stderr = FALSE
)
```

```
Pi is roughly 3.1415503141550314
```

If the preceding message is not displayed, you will need to investigate why your Spark cluster is not properly configured, which is beyond the scope of this book. As a start, rerun the Pi example but remove `stderr = FALSE`; this prints errors to the console, which you then can use to investigate what the problem might be. When using a cloud provider or a Spark distribution, you can contact their support team to help you troubleshoot this further; otherwise, Stack Overflow is a good place to start.

If you do see the message, this means that your Spark cluster is properly configured but somehow R is not able to use Spark, so you need to troubleshoot in detail, as we will explain next.

Detailed troubleshooting

To troubleshoot the connection process in detail, you can manually replicate the two-step connection process, which is often very helpful to diagnose connection issues. First, `spark-submit` is triggered from R, which submits the application to Spark; second, R connects to the running Spark application.

First, identify the Spark installation directory and the path to the correct `spar klyr*.jar` file by running the following:

```
dir(system.file("java", package = "sparklyr"),
    pattern = "sparklyr", full.names = T)
```

Ensure that you identify the correct version that matches your Spark cluster—for instance, `sparklyr-2.1-2.11.jar` for Spark 2.1.

Then, from the terminal, run this:

```
$SPARK_HOME/bin/spark-submit --class sparklyr.Shell $PATH_TO_SPARKLYR_JAR 8880 42

18/06/11 12:13:53 INFO sparklyr: Session (42) found port 8880 is available
18/06/11 12:13:53 INFO sparklyr: Gateway (42) is waiting for sparklyr client
                                 to connect to port 8880
```

The parameter 8880 represents the default port to use in `sparklyr`, while 42 is the session number, which is a cryptographically secure number generated by `sparklyr`, but for troubleshooting purposes can be as simple as 42.

If this first connection step fails, it means that the cluster can't accept the application. This usually means that there are not enough resources, or there are permission restrictions.

The second step is to connect from R as follows (notice that there is a 60-second timeout, so you'll need to run the R command after running the terminal command; if needed, you can configure this timeout as described in Chapter 9):

```
library(sparklyr)
sc <- spark_connect(master = "sparklyr://localhost:8880/42", version = "2.3")
```

If this second connection step fails, it usually means that there is a connectivity problem between R and the driver node. You can try using a different connection port, for instance.

Windows

Connecting from Windows is, in most cases, as straightforward as connecting from Linux and macOS. However, there are a few common connection issues that you should be aware of:

- Firewalls and antivirus software might block ports for your connection. The default port used by `sparklyr` is 8880; double-check that this port is not being blocked.
- Long path names can cause issues, especially with older Windows systems like Windows 7. When you're using these systems, try connecting with Spark installed with all folders, using at most eight characters and no spaces in their names.

Recap

This chapter presented an overview of Spark's architecture, connection concepts, and examples to connect in local mode, standalone, YARN, Mesos, Kubernetes, and Livy. It also presented edge nodes and their role while connecting to Spark clusters. This should have provided you with enough information to successfully connect to any Apache Spark cluster.

To troubleshoot connection problems beyond the techniques described in this chapter, we recommend that you search for the connection problem in Stack Overflow, the `sparklyr` issues GitHub page (*http://bit.ly/2Z72XWa*), and, if needed, open a new GitHub issue in `sparklyr` (*http://bit.ly/2HasCmq*) to assist further.

In Chapter 8, we cover how to use Spark to read and write from a variety of data sources and formats, which allows you to be more agile when adding new data sources for data analysis. What used to take days, weeks, or even months, you now can complete in hours by embracing data lakes.

Data

*Has it occurred to you that she might not have
been a reliable source of information?*

—*Jon Snow*

With the knowledge acquired in previous chapters, you are now equipped to start doing analysis and modeling at scale! So far, however, we haven't really explained much about how to read data into Spark. We've explored how to use `copy_to()` to upload small datasets or functions like `spark_read_csv()` or `spark_write_csv()` without explaining in detail how and why.

So, you are about to learn how to read and write data using Spark. And, while this is important on its own, this chapter will also introduce you to the *data lake*—a repository of data stored in its natural or raw format that provides various benefits over existing storage architectures. For instance, you can easily integrate data from external systems without transforming it into a common format and without assuming those sources are as reliable as your internal data sources.

In addition, we will also discuss how to extend Spark's capabilities to work with data not accessible out of the box and make several recommendations focused on improving performance for reading and writing data. Reading large datasets often requires you to fine-tune your Spark cluster configuration, but that's the topic of Chapter 9.

Overview

In Chapter 1, you learned that beyond big data and big compute, you can also use Spark to improve velocity, variety, and veracity in data tasks. While you can use the learnings of this chapter for any task requiring loading and storing data, it is particularly interesting to present this chapter in the context of dealing with a variety of data sources. To understand why, we should first take a quick detour to examine how data is currently processed in many organizations.

For several years, it's been a common practice to store large datasets in a relational *database*, a system first proposed in 1970 by Edgar F. Codd.[1] You can think of a database as a collection of tables that are related to one another, where each table is carefully designed to hold specific data types and relationships to other tables. Most relational database systems use *Structured Query Language* (SQL) for querying and maintaining the database. Databases are still widely used today, with good reason: they store data reliably and consistently; in fact, your bank probably stores account balances in a database and that's a good practice.

However, databases have also been used to store information from other applications and systems. For instance, your bank might also store data produced by other banks, such as incoming checks. To accomplish this, the external data needs to be extracted from the external system, transformed into something that fits the current database, and finally be loaded into it. This is known as *Extract, Transform, and Load* (ETL), a general procedure for copying data from one or more sources into a destination system that represents the data differently from the source. The ETL process became popular in the 1970s.

Aside from databases, data is often also loaded into a *data warehouse*, a system used for reporting and data analysis. The data is usually stored and indexed in a format that increases data analysis speed but that is often not suitable for modeling or running custom distributed code. The challenge is that changing databases and data warehouses is usually a long and delicate process, since data needs to be reindexed and the data from multiple data sources needs to be carefully transformed into single tables that are shared across data sources.

Instead of trying to transform all data sources into a common format, you can embrace this variety of data sources in a *data lake*, a system or repository of data stored in its natural format (see Figure 8-1). Since data lakes make data available in its original format, there is no need to carefully transform it in advance; anyone can use it for analysis, which adds significant flexibility over ETL. You then can use Spark to unify data processing from data lakes, databases, and data warehouses through a single interface that is scalable across all of them. Some organizations also use Spark to replace their existing ETL process; however, this falls in the realm of data engineering, which is well beyond the scope of this book. We illustrate this with dotted lines in Figure 8-1.

---

1 Codd EF (1970). "A relational model of data for large shared data banks."

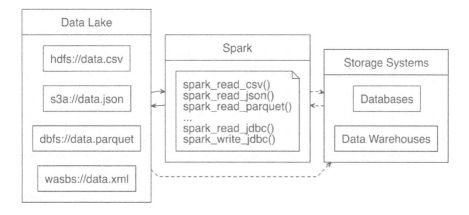

Figure 8-1. Spark processing raw data from a data lakes, databases, and data warehouses

To support a broad variety of data sources, Spark needs to be able to read and write data in several different file formats (CSV, JSON, Parquet, and others), and access them while stored in several file systems (HDFS, S3, DBFS, and more) and, potentially, interoperate with other storage systems (databases, data warehouses, etc.). We will get to all of that, but first, we will start by presenting how to read, write, and copy data using Spark.

Reading Data

If you are new to Spark, it is highly recommended to review this section before you start working with large datasets. We will introduce several techniques that improve the speed and efficiency of reading data. Each subsection presents specific ways to take advantage of how Spark reads files, such as the ability to treat entire folders as datasets as well as being able to describe them to read datasets faster in Spark.

Paths

When analyzing data, loading multiple files into a single data object is a common scenario. In R, we typically use a loop or a functional programming directive to accomplish this. This is because R must load each file individually into your R session. Let's create a few CSV files inside a folder and read them with R first:

```
letters <- data.frame(x = letters, y = 1:length(letters))

dir.create("data-csv")
write.csv(letters[1:3, ], "data-csv/letters1.csv", row.names = FALSE)
write.csv(letters[1:3, ], "data-csv/letters2.csv", row.names = FALSE)
```

```
do.call("rbind", lapply(dir("data-csv", full.names = TRUE), read.csv))
```

```
  x y
1 a 1
2 b 2
3 c 3
4 a 1
5 b 2
6 c 3
```

In Spark, there is the notion of a folder as a dataset. Instead of enumerating each file, simply pass the path containing all the files. Spark assumes that every file in that folder is part of the same dataset. This implies that the target folder should be used only for data purposes. This is especially important since storage systems like HDFS store files across multiple machines, but, conceptually, they are stored in the same folder; when Spark reads the files from this folder, it's actually executing distributed code to read each file within each machine—no data is transferred between machines when distributed files are read:

```
library(sparklyr)
sc <- spark_connect(master = "local", version = "2.3")

spark_read_csv(sc, "data-csv/")
# Source: spark<datacsv> [?? x 2]
      x        y
   <chr> <int>
 1 a        1
 2 b        2
 3 c        3
 4 d        4
 5 e        5
 6 a        1
 7 b        2
 8 c        3
 9 d        4
10 e        5
```

The "folder as a table" idea is found in other open source technologies as well. Under the hood, Hive tables work the same way. When you query a Hive table, the mapping is done over multiple files within the same folder. The folder's name usually matches the name of the table visible to the user.

Next, we will present a technique that allows Spark to read files faster as well as to reduce read failures by describing the structure of a dataset in advance.

Schema

When reading data, Spark is able to determine the data source's column names and column types, also known as the *schema*. However, guessing the schema comes at a

---

cost; Spark needs to do an initial pass on the data to guess what it is. For a large data-set, this can add a significant amount of time to the data ingestion process, which can become costly even for medium-size datasets. For files that are read over and over again, the additional read time accumulates over time.

To avoid this, Spark allows you to provide a column definition by providing a col umns argument to describe your dataset. You can create this schema by sampling a small portion of the original file yourself:

```
spec_with_r <- sapply(read.csv("data-csv/letters1.csv", nrows = 10), class)
spec_with_r
       x         y
  "factor" "integer"
```

Or, you can set the column specification to a vector containing the column types explicitly. The vector's values are named to match the field names:

```
spec_explicit <- c(x = "character", y = "numeric")
spec_explicit
         x           y
 "character"   "numeric"
```

The accepted variable types are: integer, character, logical, double, numeric, `factor`, Date, and POSIXct.

Then, when reading using spark_read_csv(), you can pass spec_with_r to the col umns argument to match the names and types of the original file. This helps to improve performance since Spark will not need to determine the column types.

```
spark_read_csv(sc, "data-csv/", columns = spec_with_r)
# Source: spark<datacsv> [?? x 2]
    x     y
  <chr> <int>
1 a       1
2 b       2
3 c       3
4 a       1
5 b       2
6 c       3
```

The following example shows how to set the field type to something different. How-ever, the new field type needs to be a compatible type in the original dataset. For example, you cannot set a character field to numeric. If you use an incompatible type, the file read will fail with an error. Additionally, the following example also changes the names of the original fields:

```
spec_compatible <- c(my_letter = "character", my_number = "character")

spark_read_csv(sc, "data-csv/", columns = spec_compatible)
```

```
# Source: spark<datacsv> [?? x 2]
  my_letter my_number
  <chr>     <chr>
1 a         1
2 b         2
3 c         3
4 a         1
5 b         2
6 c         3
```

In Spark, malformed entries can cause errors during reading, particularly for non-character fields. To prevent such errors, we can use a file specification that imports them as characters and then use dplyr to coerce the field into the desired type.

This subsection reviewed how we can read files faster and with fewer failures, which lets us start our analysis more quickly. Another way to accelerate our analysis is by loading less data into Spark memory, which we examine in the next section.

Memory

By default, when using Spark with R, when you read data, it is copied into Spark's distributed memory, making data analysis and other operations very fast. There are cases, such as when the data is too big, for which loading all the data might not be practical or even necessary. For those cases, Spark can just "map" the files without copying data into memory.

The mapping creates a sort of virtual table in Spark. The implication is that when a query runs against that table, Spark needs to read the data from the files at that time. Any consecutive reads after that will do the same. In effect, Spark becomes a pass-through for the data. The advantage of this method is that there is almost no up-front time cost to "reading" the file; the mapping is very fast. The downside is that running queries that actually extract data will take longer.

This is controlled by the memory argument of the read functions. Setting it to FALSE prevents the data copy (the default is TRUE):

```
mapped_csv <- spark_read_csv(sc, "data-csv/", memory = FALSE)
```

There are good use cases for this method, one of which is when not all columns of a table are needed. For example, take a very large file that contains many columns. Assuming this is not the first time you interact with this data, you would know what columns are needed for the analysis. When you know which columns you need, the files can be read using memory = FALSE, and then the needed columns can be selected with dplyr. The resulting dplyr variable can then be cached into memory, using the compute() function. This will make Spark query the file(s), pull the selected fields, and copy only that data into memory. The result is an in-memory table that took comparatively less time to ingest:

```
mapped_csv %>%
  dplyr::select(y) %>%
  dplyr::compute("test")
```

The next section covers a short technique to make it easier to carry the original field names of imported data.

Columns

Spark 1.6 required that column names be sanitized, so R does that by default. There might be cases when you would like to keep the original names intact, or when working with Spark version 2.0 or above. To do that, set the `sparklyr.sanitize.col` `umn.names` option to FALSE:

```
options(sparklyr.sanitize.column.names = FALSE)
copy_to(sc, iris, overwrite = TRUE)

# Source:   table<iris> [?? x 5]
# Database: spark_connection
   Sepal.Length Sepal.Width Petal.Length Petal.Width Species
          <dbl>       <dbl>        <dbl>       <dbl> <chr>
1           5.1         3.5          1.4         0.2 setosa
2           4.9         3            1.4         0.2 setosa
3           4.7         3.2          1.3         0.2 setosa
4           4.6         3.1          1.5         0.2 setosa
5           5           3.6          1.4         0.2 setosa
6           5.4         3.9          1.7         0.4 setosa
7           4.6         3.4          1.4         0.3 setosa
8           5           3.4          1.5         0.2 setosa
9           4.4         2.9          1.4         0.2 setosa
10          4.9         3.1          1.5         0.1 setosa
# ... with more rows
```

With this review of how to read data into Spark, we move on to look at how we can write data from our Spark session.

Writing Data

Some projects require that new data generated in Spark be written back to a remote source. For example, the data could be new predicted values returned by a Spark model. The job processes the mass generation of predictions, but then the predictions need to be stored. This section focuses on how you should use Spark for moving the data from Spark into an external destination.

Many new users start by downloading Spark data into R, and then upload it to a target, as illustrated in Figure 8-2. It works for smaller datasets, but it becomes inefficient for larger ones. The data typically grows in size to the point that it is no longer feasible for R to be the middle point.

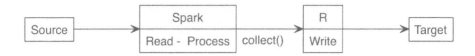

Figure 8-2. Incorrect use of Spark when writing large datasets

All efforts should be made to have Spark connect to the target location. This way, reading, processing, and writing happens within the same Spark session.

As Figure 8-3 shows, a better approach is to use Spark to read, process, and write to the target. This approach is able to scale as big as the Spark cluster allows, and prevents R from becoming a choke point.

Figure 8-3. Correct use of Spark when writing large datasets

Consider the following scenario: a Spark job just processed predictions for a large dataset, resulting in a considerable amount of predictions. Choosing a method to write results will depend on the technology infrastructure you are working on. More specifically, it will depend on Spark and the target running, or not, in the same cluster.

Back to our scenario, we have a large dataset in Spark that needs to be saved. When Spark and the target are in the same cluster, copying the results is not a problem; the data transfer is between RAM and disk of the same cluster or efficiently shuffled through a high-bandwidth connection.

But what to do if the target is not within the Spark cluster? There are two options, and choosing one will depend on the size of the data and network speed:

Spark transfer
> In this case, Spark connects to the remote target location and copies the new data. If this is done within the same datacenter, or cloud provider, the data transfer could be fast enough to have Spark write the data directly.

External transfer and otherwise
> Spark can write the results to disk and transfers them via a third-party application. Spark writes the results as files and then a separate job copies the files over. In the target location, you would use a separate process to transfer the data into the target location.

It is best to recognize that Spark, R, and any other technology are tools. No tool can do everything, nor should you expect it to. Next we describe how to copy data into Spark or collect large datasets that don't fit in memory, which you can use to transfer data across clusters or help initialize your distributed datasets.

Copying Data

Previous chapters used `copy_to()` as a handy helper to copy data into Spark; however, you can use `copy_to()` only to transfer in-memory datasets that are already loaded in memory. These datasets tend to be much smaller than the kind of datasets you would want to copy into Spark.

For instance, suppose that we have a 3 GB dataset generated as follows:

```
dir.create("largefile.txt")
write.table(matrix(rnorm(10 * 10^6), ncol = 10), "largefile.txt/1",
            append = T, col.names = F, row.names = F)
for (i in 2:30)
  file.copy("largefile.txt/1", paste("largefile.txt/", i))
```

If we had only 2 GB of memory in the driver node, we would not be able to load this 3 GB file into memory using `copy_to()`. Instead, when using the HDFS as storage in your cluster, you can use the `hadoop` command-line tool to copy files from disk into Spark from the terminal as follows. Notice that the following code works only in clusters using HDFS, not in local environments.

```
hadoop fs -copyFromLocal largefile.txt largefile.txt
```

You then can read the uploaded file, as described in the "File Formats" on page 144 section; for text files, you would run:

```
spark_read_text(sc, "largefile.txt", memory = FALSE)

# Source: spark<largefile> [?? x 1]
   line
   <chr>
 1 0.0982531064914565 -0.577567317599452 -1.66433938237253 -0.20095089489…
 2 -1.08322304504007 1.05962389624635 1.1852771207729 -0.230934710049462 …
 3 -0.398079835552421 0.293643382374479 0.727994248743204 -1.571547990532…
 4 0.418899768227183 0.534037617828835 0.921680317620166 -1.6623094393911…
 5 -0.204409401553028 -0.0376212693728992 -1.13012269711811 0.56149527218…
 6 1.41192628218417 -0.580413572014808 0.727722566256326 0.5746066486689 …
 7 -0.313975036262443 -0.0166426329807508 -0.188906975208319 -0.986203251…
 8 -0.571574679637623 0.513472254005066 0.139050812059352 -0.822738334753…
 9 1.39983023148955 -1.08723592838627 1.02517804413913 -0.412680186313667…
10 0.6318328148434 -1.08741784644221 -0.550575696474202 0.971967251067794…
# … with more rows
```

`collect()` has a similar limitation in that it can collect only datasets that fit your driver memory; however, if you had to extract a large dataset from Spark through the driver node, you could use specialized tools provided by the distributed storage. For HDFS, you would run the following:

```
hadoop fs -copyToLocal largefile.txt largefile.txt
```

Alternatively, you can also collect datasets that don't fit in memory by providing a callback to `collect()`. A callback is just an R function that will be called over each Spark partition. You then can write this dataset to disk or push to other clusters over the network.

You could use the following code to collect 3 GB even if the driver node collecting this dataset had less than 3 GB of memory. That said, as Chapter 3 explains, you should avoid collecting large datasets into a single machine since this creates a significant performance bottleneck. For conciseness, we will collect only the first million rows; feel free to remove `head(10^6)` if you have a few minutes to spare:

```
dir.create("large")
spark_read_text(sc, "largefile.txt", memory = FALSE) %>%
  head(10^6) %>%
  collect(callback = function(df, idx) {
    writeLines(df$line, paste0("large/large-", idx, ".txt"))
  })
```

Make sure you clean up these large files and empty your recycle bin as well:

```
unlink("largefile.txt", recursive = TRUE)
unlink("large", recursive = TRUE)
```

In most cases, data will already be stored in the cluster, so you should not need to worry about copying large datasets; instead, you can usually focus on reading and writing different file formats, which we describe next.

File Formats

Out of the box, Spark is able to interact with several file formats, like CSV, JSON, LIBSVM, ORC, and Parquet. Table 8-1 maps the file format to the function you should use to read and write data in Spark.

Table 8-1. Spark functions to read and write file formats

Format	Read	Write
CSV	spark_read_csv()	spark_write_csv()
JSON	spark_read_json()	spark_write_json()
LIBSVM	spark_read_libsvm()	spark_write_libsvm()
ORC	spark_read_orc()	spark_write_orc()
Apache Parquet	spark_read_parquet()	spark_write_parquet()
Text	spark_read_text()	spark_write_text()

The following sections will describe special considerations particular to each file format as well as some of the strengths and weaknesses of some popular file formats, starting with the well-known CSV file format.

CSV

The CSV format might be the most common file type in use today. It is defined by a text file separated by a given character, usually a comma. It should be pretty straightforward to read CSV files; however, it's worth mentioning a couple techniques that can help you process CSVs that are not fully compliant with a well-formed CSV file. Spark offers the following modes for addressing parsing issues:

Permissive
Inserts NULL values for missing tokens

Drop Malformed
Drops lines that are malformed

Fail Fast
Aborts if it encounters any malformed line

You can use these in `sparklyr` by passing them inside the `options` argument. The following example creates a file with a broken entry. It then shows how it can be read into Spark:

```
## Creates bad test file
writeLines(c("bad", 1, 2, 3, "broken"), "bad.csv")

spark_read_csv(
  sc,
  "bad3",
  "bad.csv",
  columns = list(foo = "integer"),
  options = list(mode = "DROPMALFORMED"))
# Source: spark<bad3> [?? x 1]
    foo
  <int>
1     1
2     2
3     3
```

Spark provides an issue tracking column, which was hidden by default. To enable it, add `_corrupt_record` to the `columns` list. You can combine this with the use of the `PERMISSIVE` mode. All rows will be imported, invalid entries will receive an `NA`, and the issue will be tracked in the `_corrupt_record` column:

```
spark_read_csv(
  sc,
  "bad2",
```

```
    "bad.csv",
    columns = list(foo = "integer", "_corrupt_record" = "character"),
    options = list(mode = "PERMISSIVE")
)
# Source: spark<bad2> [?? x 2]
    foo `_corrupt_record`
  <int> <chr>
1     1 NA
2     2 NA
3     3 NA
4    NA broken
```

Reading and storing data as CSVs is quite common and supported across most systems. For tabular datasets, it is still a popular option, but for datasets containing nested structures and nontabular data, JSON is usually preferred.

JSON

JSON is a file format originally derived from JavaScript that has grown to be language-independent and very popular due to its flexibility and ubiquitous support. Reading and writing JSON files is quite straightforward:

```
writeLines("{'a':1, 'b': {'f1': 2, 'f3': 3}}", "data.json")
simple_json <- spark_read_json(sc, "data.json")
simple_json

# Source: spark<data> [?? x 2]
      a b
  <dbl> <list>
1     1 <list [2]>
```

However, when you deal with a dataset containing nested fields like the one from this example, it is worth pointing out how to extract nested fields. One approach is to use a JSON path, which is a domain-specific syntax commonly used to extract and query JSON files. You can use a combination of `get_json_object()` and `to_json()` to specify the JSON path you are interested in. To extract `f1` you would run the following transformation:

```
simple_json %>% dplyr::transmute(z = get_json_object(to_json(b), '$.f1'))

# Source: spark<?> [?? x 3]
      a b            z
  <dbl> <list>       <chr>
1     1 <list [2]>   2
```

Another approach is to install `sparkly.nested` from CRAN with `install.pack ages("sparklyr.nested")` and then unnest nested data with `sdf_unnest()`:

```
sparklyr.nested::sdf_unnest(simple_json, "b")

# Source: spark<?> [?? x 3]
      a    f1    f3
```

```
<dbl> <dbl> <dbl>
 1    1    2    3
```

While JSON and CSVs are quite simple to use and versatile, they are not optimized for performance; however, other formats like ORC, AVRO, and Parquet are.

Parquet

Apache Parquet, Apache ORC, and Apache AVRO are all file formats designed with performance in mind. Parquet and ORC store data in columnar format, while AVRO is row-based. All of them are binary file formats, which reduces storage space and improves performance. This comes at the cost of making them a bit more difficult to read by external systems and libraries; however, this is usually not an issue when used as intermediate data storage within Spark.

To illustrate this, Figure 8-4 plots the result of running a 1-million-row write-speed benchmark using the bench package; feel free to use your own benchmarks over meaningful datasets when deciding which format best fits your needs:

```
numeric <- copy_to(sc, data.frame(nums = runif(10^6)))
bench::mark(
  CSV = spark_write_csv(numeric, "data.csv", mode = "overwrite"),
  JSON = spark_write_json(numeric, "data.json", mode = "overwrite"),
  Parquet = spark_write_parquet(numeric, "data.parquet", mode = "overwrite"),
  ORC = spark_write_parquet(numeric, "data.orc", mode = "overwrite"),
  iterations = 20
) %>% ggplot2::autoplot()
```

From now on, be sure to disconnect from Spark whenever we present a new spark_connect() command:

```
spark_disconnect(sc)
```

This concludes the introduction to some of the out-of-the-box supported file formats. Next, we describe how to deal with formats that require external packages and customization.

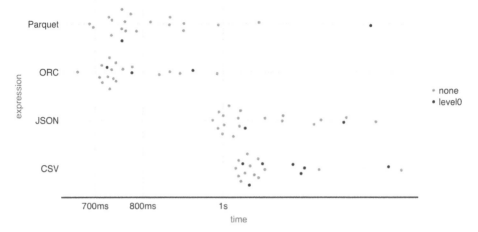

Figure 8-4. One-million-rows write benchmark between CSV, JSON, Parquet, and ORC

Others

Spark is a very flexible computing platform. It can add functionality by using extension programs, called packages. You can access a new source type or file system by using the appropriate package.

Packages need to be loaded into Spark at connection time. To load the package, Spark needs its location, which could be inside the cluster, in a file share, or the internet.

In sparklyr, the package location is passed to spark_connect(). All packages should be listed in the sparklyr.connect.packages entry of the connection configuration.

It is possible to access data source types that we didn't previously list. Loading the appropriate default package for Spark is the first of two steps The second step is to actually read or write the data. The spark_read_source() and spark_write_source() functions do that. They are generic functions that can use the libraries imported by a default package.

For instance, we can read XML files as follows:

```
sc <- spark_connect(master = "local", version = "2.3", config = list(
  sparklyr.connect.packages = "com.databricks:spark-xml_2.11:0.5.0"))

writeLines("<ROWS><ROW><text>Hello World</text></ROW>", "simple.xml")
spark_read_source(sc, "simple_xml", "simple.xml", "xml")

# Source: spark<data> [?? x 1]
  text
  <chr>
1 Hello World
```

which you can also write back to XML with ease, as follows:

```
tbl(sc, "simple_xml") %>%
  spark_write_source("xml", options = list(path = "data.xml"))
```

In addition, there are a few extensions developed by the R community to load additional file formats, such as `sparklyr.nested` to assist with nested data, `spark.sas7bdat` to read data from SAS, `sparkavro` to read data in AVRO format, and `sparkwarc` to read WARC files, which use extensibility mechanisms introduced in Chapter 10. Chapter 11 presents techniques to use R packages to load additional file formats, and Chapter 13 presents techniques to use Java libraries to complement this further. But first, let's explore how to retrieve and store files from several different file systems.

File Systems

Spark defaults to the file system on which it is currently running. In a YARN managed cluster, the default file system will be HDFS. An example path of */home/user/ file.csv* will be read from the cluster's HDFS folders, not the Linux folders. The operating system's file system will be accessed for other deployments, such as Standalone, and `sparklyr`'s local.

The file system protocol can be changed when reading or writing. You do this via the `path` argument of the `sparklyr` function. For example, a full path of *file:///home/user/ file.csv* forces the use of the local operating system's file system.

There are many other file system protocols, such as `_dbfs://_` for Databricks' file system, `_s3a://_` for Amazon's S3 service, `_wasb://_` for Microsoft Azure storage, and `_gs://_` for Google storage.

Spark does not provide support for all them directly; instead, they are configured as needed. For instance, accessing the "s3a" protocol requires adding a package to the `sparklyr.connect.packages` configuration setting, while connecting and specifying appropriate credentials might require using the `AWS_ACCESS_KEY_ID` and `AWS_SECRET_ACCESS_KEY` environment variables.

```
Sys.setenv(AWS_ACCESS_KEY_ID = my_key_id)
Sys.setenv(AWS_SECRET_ACCESS_KEY = my_secret_key)

sc <- spark_connect(master = "local", version = "2.3", config = list(
  sparklyr.connect.packages = "org.apache.hadoop:hadoop-aws:2.7.7"))

my_file <- spark_read_csv(sc, "my-file", path =  "s3a://my-bucket/my-file.csv")
```

Accessing other file protocols requires loading different packages, although, in some cases, the vendor providing the Spark environment might load the package for you. Please refer to your vendor's documentation to find out whether that is the case.

Storage Systems

A data lake and Spark usually go hand-in-hand, with optional access to storage systems like databases and data warehouses. Presenting all the different storage systems with appropriate examples would be quite time-consuming, so instead we present some of the commonly used storage systems.

As a start, Apache *Hive* is a data warehouse software that facilitates reading, writing, and managing large datasets residing in distributed storage using SQL. In fact, Spark has components from Hive built directly into its sources. It is very common to have installations of Spark or Hive side-by-side, so we will start by presenting Hive, followed by Cassandra, and then close by looking at JDBC connections.

Hive

In YARN managed clusters, Spark provides a deeper integration with Apache Hive. Hive tables are easily accessible after opening a Spark connection.

You can access a Hive table's data using DBI by referencing the table in a SQL statement:

```
sc <- spark_connect(master = "local", version = "2.3")
spark_read_csv(sc, "test", "data-csv/", memory = FALSE)

DBI::dbGetQuery(sc, "SELECT * FROM test limit 10")
```

Another way to reference a table is with dplyr using the tbl() function, which retrieves a reference to the table:

```
dplyr::tbl(sc, "test")
```

It is important to reiterate that no data is imported into R; the tbl() function only creates a reference. You then can pipe more dplyr verbs following the tbl() command:

```
dplyr::tbl(sc, "test") %>%
  dplyr::group_by(y) %>%
  dplyr::summarise(totals = sum(y))
```

Hive table references assume a default database source. Often, the needed table is in a different database within the metastore. To access it using SQL, prefix the database name to the table. Separate them using a period, as demonstrated here:

```
DBI::dbSendQuery(sc, "SELECT * FROM databasename.table")
```

In dplyr, the in_schema() function can be used. The function is used inside the tbl() call:

```
tbl(sc, dbplyr::in_schema("databasename", "table"))
```

You can also use the `tbl_change_db()` function to set the current session's default database. Any subsequent call via `DBI` or `dplyr` will use the selected name as the default database:

```
tbl_change_db(sc, "databasename")
```

The following examples require additional Spark packages and databases which might be difficult to follow unless you happen to have a JDBC driver or Cassandra database accessible to you.

Next, we explore a less structured storage system, often referred to as a *NoSQL database*.

Cassandra

Apache *Cassandra* is a free and open source, distributed, wide-column store, NoSQL database management system designed to handle large amounts of data across many commodity servers. While there are many other database systems beyond Cassandra, taking a quick look at how Cassandra can be used from Spark will give you insight into how to make use of other database and storage systems like Solr, Redshift, Delta Lake, and others.

The following example code shows how to use the `datastax:spark-cassandra-connector` package to read from Cassandra. The key is to use the `org.apache.spark.sql.cassandra` library as the `source` argument. It provides the mapping Spark can use to make sense of the data source. Unless you have a Cassandra database, skip executing the following statement:

```
sc <- spark_connect(master = "local", version = "2.3", config = list(
  sparklyr.connect.packages = "datastax:spark-cassandra-connector:2.3.1-s_2.11"))

spark_read_source(
  sc,
  name = "emp",
  source = "org.apache.spark.sql.cassandra",
  options = list(keyspace = "dev", table = "emp"),
  memory = FALSE)
```

One of the most useful features of Spark when dealing with external databases and data warehouses is that Spark can push down computation to the database, a feature known as *pushdown predicates*. In a nutshell, pushdown predicates improve performance by asking remote databases smart questions. When you execute a query that contains the `filter(age > 20)` expression against a remote table referenced through `spark_read_source()` and not loaded in memory, rather than bringing the entire table into Spark, it will be passed to the remote database and only a subset of the remote table is retrieved.

While it is ideal to find Spark packages that support the remote storage system, there will be times when a package is not available and you need to consider vendor JDBC drivers.

JDBC

When a Spark package is not available to provide connectivity, you can consider a JDBC connection. JDBC is an interface for the programming language Java, which defines how a client can access a database.

It is quite easy to connect to a remote database with spark_read_jdbc(), and spark_write_jdbc(); as long as you have access to the appropriate JDBC driver, which at times is trivial and other times is quite an adventure. To keep this simple, we can briefly consider how a connection to a remote MySQL database could be accomplished.

First, you would need to download the appropriate JDBC driver from MySQL's developer portal and specify this additional driver as a sparklyr.shell.driver-class-path connection option. Since JDBC drivers are Java-based, the code is contained within a *JAR* (Java ARchive) file. As soon as you're connected to Spark with the appropriate driver, you can use the *jdbc://* protocol to access particular drivers and databases. Unless you are willing to download and configure MySQL on your own, skip executing the following statement:

```
sc <- spark_connect(master = "local", version = "2.3", config = list(
  "sparklyr.shell.driver-class-path" =
    "~/Downloads/mysql-connector-java-5.1.41/mysql-connector-java-5.1.41-bin.jar"
))

spark_read_jdbc(sc, "person_jdbc",  options = list(
  url = "jdbc:mysql://localhost:3306/sparklyr",
  user = "root", password = "<password>",
  dbtable = "person"))
```

If you are a customer of particular database vendors, making use of the vendor-provided resources is usually the best place to start looking for appropriate drivers.

Recap

This chapter expanded on how and why you should use Spark to connect and process a variety of data sources through a new data storage model known as data lakes—a storage pattern that provides more flexibility than standard ETL processes by enabling you to use raw datasets with, potentially, more information to enrich data analysis and modeling.

We also presented best practices for reading, writing, and copying data into and from Spark. We then returned to exploring the components of a data lake: file formats and file systems, with the former representing how data is stored, and the latter where the data is stored. You then learned how to tackle file formats and storage systems that require additional Spark packages, reviewed some of the performance trade-offs across file formats, and learned the concepts required to make use of storage systems (databases and warehouses) in Spark.

While reading and writing datasets should come naturally to you, you might still hit resource restrictions while reading and writing large datasets. To handle these situations, Chapter 9 shows you how Spark manages tasks and data across multiple machines, which in turn allows you to further improve the performance of your analysis and modeling tasks.

Tuning

Chaos isn't a pit. Chaos is a ladder.

—*Petyr Baelish*

In previous chapters, we've assumed that computation within a Spark cluster works efficiently. While this is true in some cases, it is often necessary to have some knowledge of the operations Spark runs internally to fine-tune configuration settings that will make computations run efficiently. This chapter explains how Spark computes data over large datasets and provides details on how to optimize its operations.

For instance, in this chapter you'll learn how to request more compute nodes and increase the amount of memory, which, if you remember from Chapter 2, defaults to only 2 GB in local instances. You will learn how Spark unifies computation through partitioning, shuffling, and caching. As mentioned a few chapters back, this is the last chapter describing the internals of Spark; after you complete this chapter, we believe that you will have the intermediate Spark skills necessary to be productive at using Spark.

In Chapters 10–12 we explore exciting techniques to deal with specific modeling, scaling, and computation problems. However, we must first understand how Spark performs internal computations, what pieces we can control, and why.

Overview

Spark performs distributed computation by configuring, partitioning, executing, shuffling, caching, and serializing data, tasks, and resources across multiple machines:

- *Configuring* requests the cluster manager for resources: total machines, memory, and so on.

- *Partitioning* splits the data among various machines. Partitions can be either implicit or explicit.

- *Executing* means running an arbitrary transformation over each partition.

- *Shuffling* redistributes data to the correct machine.

- *Caching* preserves data in memory across different computation cycles.

- *Serializing* transforms data to be sent over the network to other workers or back to the driver node.

To illustrate each concept, let's create three partitions with unordered integers and then sort them using `arrange()`:

```
data <- copy_to(sc,
  data.frame(id = c(4, 9, 1, 8, 2, 3, 5, 7, 6)),
  repartition = 3)
```

```
data %>% arrange(id) %>% collect()
```

Figure 9-1 shows how this sorting *job* would conceptually work across a cluster of machines. First, Spark would *configure* the cluster to use three worker machines. In this example, the numbers 1 through 9 are partitioned across three storage instances. Since the *data* is already partitioned, each worker node loads this implicit *partition*; for instance, 4, 9, and 1 are loaded in the first worker node. Afterward, a *task* is distributed to each worker to apply a transformation to each data partition in each worker node; this task is denoted by f(x). In this example, f(x) *executes* a sorting operation within a partition. Since Spark is general, execution over a partition can be as simple or complex as needed.

The result is then *shuffled* to the correct machine to finish the sorting operation across the entire dataset, which completes a stage. A *stage* is a set of operations that Spark can execute without shuffling data between machines. After the data is sorted across the cluster, the sorted results can be optionally *cached* in memory to avoid rerunning this computation multiple times.

Finally, a small subset of the results is *serialized*, through the network connecting the cluster machines, back to the driver node to print a preview of this sorting example.

Notice that while Figure 9-1 describes a sorting operation, a similar approach applies to filtering or joining datasets and analyzing and modeling data at scale. Spark provides support to perform custom partitions, custom shuffling, and so on, but most of these lower-level operations are not exposed in sparklyr; instead, sparklyr makes those operations available through higher-level commands provided by data analysis tools like dplyr or DBI, modeling, and by using many extensions. For those few cases in which you might need to implement low-level operations, you can always use Spark's Scala API through sparklyr extensions or run custom distributed R code.

To effectively tune Spark, we will start by getting familiar with Spark's computation *graph* and Spark's event *timeline*. Both are accessible through Spark's web interface.

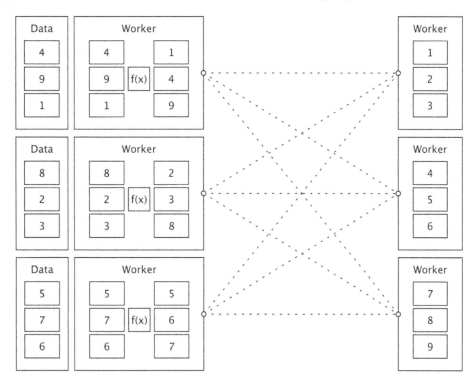

Figure 9-1. Sorting distributed data with Apache Spark

Graph

Spark describes all computation steps using a Directed Acyclic Graph (DAG), which means that all computations in Spark move computation forward without repeating previous steps, which helps Spark optimize computations effectively.

The best way to understand Spark's computation graph for a given operation—sorting for our example—is to open the last *completed query* on the SQL tab in Spark's web interface. Figure 9-2 shows the resulting graph for this sorting operation, which contains the following operations:

WholeStageCodegen

This block specifies that the operations it contains were used to generate computer code that was efficiently translated to byte code. There is usually a small cost associated with translating operations into byte code, but this is a worthwhile price to pay since the operations then can be executed much faster from Spark. In general, you can ignore this block and focus on the operations that it contains.

`InMemoryTableScan`

This means that the original dataset `data` was stored in memory and traversed row by row once.

`Exchange`

Partitions were exchanged—that is, shuffled—across executors in your cluster.

`Sort`

After the records arrived at the appropriate executor, they were sorted in this final stage.

Figure 9-2. Spark graph for a sorting query

From the query details, you then can open the last Spark job to arrive to the job details page, which you can expand by using "DAG Visualization" to create a graph similar to Figure 9-3. This graph shows a few additional details and the stages in this job. Notice that there are no arrows pointing back to previous steps, since Spark makes use of acyclic graphs.

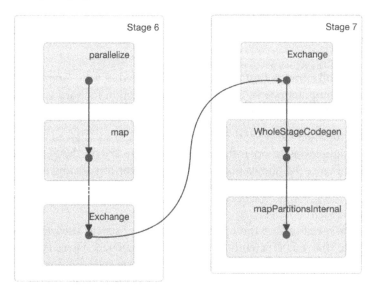

Figure 9-3. Spark graph for a sorting job

Next, we dive into a Spark stage and explore its event timeline.

Timeline

The *event timeline* is a great summary of how Spark is spending computation cycles over each stage. Ideally, you want to see this timeline consisting of mostly CPU usage since other tasks can be considered overhead. You also want to see Spark using all the CPUs across all the cluster nodes available to you.

Select the first stage in the current job and expand the event timeline, which should look similar to Figure 9-4. Notice that we explicitly requested three partitions, which are represented by three lanes in this visualization.

Figure 9-4. Spark event timeline

Since our machine is equipped with four CPUs, we can parallelize this computation even further by explicitly repartitioning data using `sdf_repartition()`, with the result shown in Figure 9-5:

```
data %>% sdf_repartition(4) %>% arrange(id) %>% collect()
```

Figure 9-5. Spark event timeline with additional partitions

Figure 9-5 now shows four execution lanes with most time spent under Executor Computing Time, which shows us that this particular operation is making better use of our compute resources. When you are working with clusters, requesting more compute nodes from your cluster should shorten computation time. In contrast, for timelines that show significant time spent shuffling, requesting more compute nodes might not shorten time and might actually make everything slower. There is no concrete set of rules to follow to optimize a stage; however, as you gain experience understanding this timeline over multiple operations, you will develop insights as to how to properly optimize Spark operations.

Configuring

When tuning a Spark application, the most common resources to configure are memory and cores, specifically:

Memory in driver
> The amount of memory required in the driver node

Memory per worker
> The amount of memory required in the worker nodes

Cores per worker
> The number of CPUs required in the worker nodes

Number of workers
> The number of workers required for this session

 It is recommended to request significantly more memory for the driver than the memory available over each worker node. In most cases, you will want to request one core per worker.

In local mode there are no workers, but we can still configure memory and cores to use through the following:

```
# Initialize configuration with defaults
config <- spark_config()

# Memory
config["sparklyr.shell.driver-memory"] <- "2g"

# Cores
config["sparklyr.connect.cores.local"] <- 2

# Connect to local cluster with custom configuration
sc <- spark_connect(master = "local", config = config)
```

When using the Spark Standalone and the Mesos cluster managers, all the available memory and cores are assigned by default; therefore, there are no additional configuration changes required, unless you want to restrict resources to allow multiple users to share this cluster. In this case, you can use `total-executor-cores` to restrict the total executors requested. The *Spark Standalone* (*http://bit.ly/307YtM6*) and *Spark on Mesos* (*http://bit.ly/31H4LCT*) guides provide additional information on sharing clusters.

When running under YARN Client, you would configure memory and cores as follows:

```
# Memory in driver
config["sparklyr.shell.driver-memory"] <- "2g"

# Memory per worker
config["sparklyr.shell.executor-memory"] <- "2g"

# Cores per worker
config["sparklyr.shell.executor-cores"] <- 1

# Number of workers
config["sparklyr.shell.num-executors"] <- 3
```

When using YARN in cluster mode you can use `sparklyr.shell.driver-cores` to configure total cores requested in the driver node. The Spark on YARN (*http://bit.ly/ 306WsQx*) guide provides additional configuration settings that can benefit you.

There are a few types of configuration settings:

Connect
> These settings are set as parameters to `spark_connect()`. They are common settings used while connecting.

Submit
> These settings are set while `sparklyr` is being submitted to Spark through `spark-submit`; some are dependent on the cluster manager being used.

Runtime
> These settings configure Spark when the Spark session is created. They are independent of the cluster manager and specific to Spark.

sparklyr
> Use these to configure `sparklyr` behavior. These settings are independent of the cluster manager and particular to R.

The following subsections present extensive lists of all the available settings. It is not required that you fully understand them all while tuning Spark, but skimming through them could prove useful in the future for troubleshooting issues. If you prefer, you can skip these subsections and use them instead as reference material as needed.

Connect Settings

You can use the parameters listed in Table 9-1 with `spark_connect()`. They configure high-level settings that define the connection method, Spark's installation path, and the version of Spark to use.

Table 9-1. Parameters used when connecting to Spark

Name	Value
master	Spark cluster URL to connect to. Use "local" to connect to a local instance of Spark installed via spark_install().
SPARK_HOME	The path to a Spark installation. Defaults to the path provided by the SPARK_HOME environment variable. If SPARK_HOME is defined, it will always be used unless the version parameter is specified to force the use of a locally installed version.
method	The method used to connect to Spark. Default connection method is "shell" to connect using spark-submit. Use "livy" to perform remote connections using HTTP, or "databricks" when using a Databricks cluster, or "qubole" when using a Qubole cluster.
app_name	The application name to be used while running in the Spark cluster.
version	The version of Spark to use. This is applicable only to "local" and "livy" connections.
config	Custom configuration for the generated Spark connection. See spark_config for details.

You can configure additional settings by specifying a list in the config parameter. Let's now take a look at what those settings can be.

Submit Settings

Some settings must be specified when spark-submit (the terminal application that launches Spark) is run. For instance, since spark-submit launches a driver node that runs as a Java instance, how much memory is allocated needs to be specified as a parameter to spark-submit.

You can list all the available spark-submit parameters by running the following:

```
spark_home_dir() %>% file.path("bin", "spark-submit") %>% system2()
```

For readability, we've provided the output of this command in Table 9-2, replacing the spark-submit parameter with the appropriate spark_config() setting and removing the parameters that are not applicable or already presented in this chapter.

Table 9-2. Setting available to configure spark-submit

Name	Value
sparklyr.shell.jars	Specified as jars parameter in spark_connect().
sparklyr.shell.packages	Comma-separated list of Maven coordinates of JARs to include on the driver and executor classpaths. Will search the local Maven repo, then Maven Central and any additional remote repositories given by sparklyr.shell.repositories. The format for the coordinates should be *groupId:artifactId:version*.
sparklyr.shell.exclude-packages	Comma-separated list of *groupId:artifactId*, to exclude while resolving the dependencies provided in sparklyr.shell.packages to avoid dependency conflicts.

Name	Value
`sparklyr.shell.repositories`	Comma-separated list of additional remote repositories to search for the Maven coordinates given with `sparklyr.shell.packages`.
`sparklyr.shell.files`	Comma-separated list of files to be placed in the working directory of each executor. Filepaths of these files in executors can be accessed via `Spark Files.get(fileName)`.
`sparklyr.shell.conf`	Arbitrary Spark configuration property set as PROP=VALUE.
`sparklyr.shell.properties-file`	Path to a file from which to load extra properties. If not specified, this will look for *conf/spark-defaults.conf*.
`sparklyr.shell.driver-java-options`	Extra Java options to pass to the driver.
`sparklyr.shell.driver-library-path`	Extra library path entries to pass to the driver.
`sparklyr.shell.driver-class-path`	Extra class path entries to pass to the driver. Note that JARs added with `sparklyr.shell.jars` are automatically included in the classpath.
`sparklyr.shell.proxy-user`	User to impersonate when submitting the application. This argument does not work with `sparklyr.shell.principal/sparklyr.shell.keytab`.
`sparklyr.shell.verbose`	Print additional debug output.

The remaining settings, shown in Table 9-3, are specific to YARN.

Table 9-3. Settings avalable to configure spark-submit when using YARN

Name	Value
`sparklyr.shell.queue`	The YARN queue to submit to (Default: "default").
`sparklyr.shell.archives`	Comma-separated list of archives to be extracted into the working directory of each executor.
`sparklyr.shell.principal`	Principal to be used to log in to KDC while running on secure HDFS.
`sparklyr.shell.keytab`	The full path to the file that contains the keytab for the principal just specified. This keytab will be copied to the node running the Application Master via the Secure Distributed Cache, for renewing the login tickets and the delegation tokens periodically.

In general, any `spark-submit` setting is configured through `sparklyr.shell.X`, where X is the name of the `spark-submit` parameter without the `--` prefix.

Runtime Settings

As mentioned, some Spark settings configure the session runtime. The runtime settings are a superset of the submit settings given that it is usually helpful to retrieve the current configuration even if a setting can't be changed.

To list the Spark settings set in your current Spark session, you can run the following:

```
spark_session_config(sc)
```

Table 9-4 describes the runtime settings.

Table 9-4. Setting available to configure the Spark session

Name	Value
spark.master	local[4]
spark.sql.shuffle.partitions	4
spark.driver.port	62314
spark.submit.deployMode	client
spark.executor.id	driver
spark.jars	/Library/.../sparklyr/java/sparklyr-2.3-2.11.jar
spark.app.id	local-1545518234395
spark.env.SPARK_LOCAL_IP	127.0.0.1
spark.sql.catalogImplementation	hive
spark.spark.port.maxRetries	128
spark.app.name	sparklyr
spark.home	/Users/.../spark/spark-2.3.2-bin-hadoop2.7
spark.driver.host	localhost

However, there are many more configuration settings available in Spark, as described in the *Spark Configuration* (*http://bit.ly/2P0Yalf*) guide. It's beyond the scope of this book to describe them all, so, if possible, take some time to identify the ones that might be of interest to your particular use cases.

sparklyr Settings

Apart from Spark settings, there are a few settings particular to sparklyr. You usually don't use these settings while tuning Spark; instead, they are helpful while trouble-shooting Spark from R. For instance, you can use sparklyr.log.console = TRUE to output the Spark logs into the R console; this is ideal while troubleshooting but too noisy otherwise. Here's how to list the settings (results are presented in Table 9-5):

```
spark_config_settings()
```

Table 9-5. Settings available to configure the sparklyr package

Name	Description
sparklyr.apply.packages	Configures default value for packages parameter in spark_apply().
sparklyr.apply.rlang	Experimental feature. Turns on improved serialization for spark_apply().

Name	Description
sparklyr.apply.serializer	Configures the version `spark_apply()` uses to serialize the closure.
sparklyr.apply.schema.infer	Number of rows collected to infer schema when column types specified in `spark_apply()`.
sparklyr.arrow	Use Apache Arrow to serialize data?
sparklyr.backend.interval	Total seconds `sparklyr` will check on a backend operation.
sparklyr.backend.timeout	Total seconds before `sparklyr` will give up waiting for a backend operation to complete.
sparklyr.collect.batch	Total rows to collect when using batch collection; defaults to 100,000.
sparklyr.cancellable	Cancel Spark jobs when the R session is interrupted?
sparklyr.connect.aftersubmit	R function to call after `spark-submit` executes.
sparklyr.connect.app.jar	The path to the `sparklyr` JAR used in `spark_connect()`.
sparklyr.connect.cores.local	Number of cores to use in `spark_connect(master = "local")`, defaults to `parallel::detectCores()`.
sparklyr.connect.csv.embedded	Regular expression to match against versions of Spark that require package extension to support CSVs.
sparklyr.connect.csv.scala11	Use Scala 2.11 JARs when using embedded CSV JARS in Spark 1.6.X.
sparklyr.connect.jars	Additional JARs to include while submitting application to Spark.
sparklyr.connect.master	The cluster master as `spark_connect()` master parameter; note that the `spark.master` setting is usually preferred.
sparklyr.connect.packages	Spark packages to include when connecting to Spark.
sparklyr.connect.ondisconnect	R function to call after `spark_disconnect()`.
sparklyr.connect.sparksubmit	Command executed instead of `spark-submit` when connecting.
sparklyr.connect.timeout	Total seconds before giving up connecting to the `sparklyr` gateway while initializing.
sparklyr.dplyr.period.splits	Should `dplyr` split column names into database and table?
sparklyr.extensions.catalog	Catalog PATH where extension JARs are located. Defaults to TRUE; FALSE to disable.
sparklyr.gateway.address	The address of the driver machine.
sparklyr.gateway.config.retries	Number of retries to retrieve port and address from config; useful when using functions to query port or address in Kubernetes.

Name	Description
sparklyr.gateway.interval	Total of seconds sparkyr will check on a gateway connection.
sparklyr.gateway.port	The port the sparklyr gateway uses in the driver machine.
sparklyr.gateway.remote	Should the sparklyr gateway allow remote connections? This is required in YARN cluster, for example.
sparklyr.gateway.routing	Should the sparklyr gateway service route to other sessions? Consider disabling in Kubernetes.
sparklyr.gateway.service	Should the sparklyr gateway be run as a service without shutting down when the last connection disconnects?
sparklyr.gateway.timeout	Total seconds before giving up connecting to the sparklyr gateway after initialization.
sparklyr.gateway.wait	Total seconds to wait before retrying to contact the sparklyr gateway.
sparklyr.livy.auth	Authentication method for Livy connections.
sparklyr.livy.headers	Additional HTTP headers for Livy connections.
sparklyr.livy.sources	Should sparklyr sources be sourced when connecting? If false, manually register sparklyr JARs.
sparklyr.log.invoke	Should every call to invoke() be printed in the console? Can be set to callstack to log call stack.
sparklyr.log.console	Should driver logs be printed in the console?
sparklyr.progress	Should job progress be reported to RStudio?
sparklyr.progress.interval	Total of seconds to wait before attempting to retrieve job progress in Spark.
sparklyr.sanitize.column.names	Should partially unsupported column names be cleaned up?
sparklyr.stream.collect.timeout	Total seconds before stopping collecting a stream sample in sdf_collect_stream().
sparklyr.stream.validate.timeout	Total seconds before stopping to check if stream has errors while being created.
sparklyr.verbose	Use verbose logging across all sparklyr operations?
sparklyr.verbose.na	Use verbose logging when dealing with NAs?
sparklyr.verbose.sanitize	Use verbose logging while sanitizing columns and other objects?
sparklyr.web.spark	The URL to Spark's web interface.
sparklyr.web.yarn	The URL to YARN's web interface.
sparklyr.worker.gateway.address	The address of the worker machine, most likely localhost.
sparklyr.worker.gateway.port	The port the sparklyr gateway uses in the driver machine.

Name	Description
`sparklyr.yarn.cluster.accepted.timeout`	Total seconds before giving up waiting for cluster resources in YARN cluster mode.
`sparklyr.yarn.cluster.hostaddress.timeout`	Total seconds before giving up waiting for the cluster to assign a host address in YARN cluster mode.
`sparklyr.yarn.cluster.lookup.byname`	Should the current username be used to filter YARN cluster jobs while searching for submitted one?
`sparklyr.yarn.cluster.lookup.prefix`	Application name prefix used to filter YARN cluster jobs while searching for submitted one.
`sparklyr.yarn.cluster.lookup.username`	The username used to filter YARN cluster jobs while searching for submitted one.
`sparklyr.yarn.cluster.start.timeout`	Total seconds before giving up waiting for YARN cluster application to get registered.

Partitioning

As mentioned in Chapter 1, MapReduce and Spark were designed with the purpose of performing computations against data stored across many machines. The subset of the data available for computation over each compute instance is known as a *partition*.

By default, Spark computes over each existing *implicit* partition since it's more effective to run computations where the data is already located. However, there are cases for which you will want to set an *explicit* partition to help Spark make more efficient use of your cluster resources.

Implicit Partitions

As Chapter 8 explained, Spark can read data stored in many formats and different storage systems; however, since shuffling data is an expensive operation, Spark executes tasks reusing the partitions in the storage system. Therefore, these partitions are implicit to Spark since they are already well defined and expensive to rearrange.

There is always an implicit partition for every computation in Spark defined by the distributed storage system, even for operations which you wouldn't expect that create partitions, like `copy_to()`.

You can explore the number of partitions a computation will require by using `sdf_num_partitions()`:

```
sdf_len(sc, 10) %>% sdf_num_partitions()

[1] 2
```

While in most cases the default partitions work just fine, there are cases for which you will need to be explicit about the partitions you choose.

Explicit Partitions

There will be times when you have many more or far fewer compute instances than data partitions. In both cases, it can help to *repartition* data to match your cluster resources.

Various data functions, like spark_read_csv(), already support a repartition parameter to request that Spark repartition data appropriately. For instance, we can create a sequence of 10 numbers partitioned by 10 as follows:

```
sdf_len(sc, 10, repartition = 10) %>% sdf_num_partitions()

[1] 10
```

For datasets that are already partitioned, we can also use sdf_repartition():

```
sdf_len(sc, 10, repartition = 10) %>%
  sdf_repartition(4) %>%
  sdf_num_partitions()

[1] 4
```

The number of partitions usually significantly changes the speed and resources being used; for instance, the following example calculates the mean over 10 million rows with different partition sizes:

```
library(microbenchmark)
library(ggplot2)

microbenchmark(
    "1 Partition(s)" = sdf_len(sc, 10^7, repartition = 1) %>%
      summarise(mean(id)) %>% collect(),
    "2 Partition(s)" = sdf_len(sc, 10^7, repartition = 2) %>%
      summarise(mean(id)) %>% collect(),
    times = 10
) %>% autoplot() + theme_light()
```

Figure 9-6 shows that sorting data with two partitions is almost twice as fast. This is because two CPUs can be used to execute this operation. However, it is not necessarily the case that higher partitions produce faster computation; instead, partitioning data is particular to your computing cluster and the data analysis operations being performed.

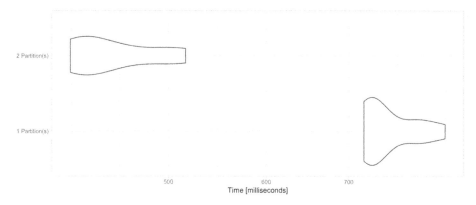

2 Partition(s)

1 Partition(s)

500   600   700
                      Time [milliseconds]

Figure 9-6. Computation speed with additional explicit partitions

Caching

Recall from Chapter 1 that Spark was designed to be faster than its predecessors by using memory instead of disk to store data. This is formally known as a Spark *resilient distributed dataset* (RDD). An RDD distributes copies of the same data across many machines, such that if one machine fails, others can complete the task—hence, the term "resilient." Resiliency is important in distributed systems since, while things will usually work in one machine, when running over thousands of machines the likelihood of something failing is much higher. When a failure happens, it is preferable to be fault tolerant to avoid losing the work of all the other machines. RDDs accomplish this by tracking data lineage information to rebuild lost data automatically on failure.

In sparklyr, you can control when an RDD is loaded or unloaded from memory using tbl_cache() and tbl_uncache().

Most sparklyr operations that retrieve a Spark DataFrame cache the results in memory. For instance, running spark_read_parquet() or copy_to() will provide a Spark DataFrame that is already cached in memory. As a Spark DataFrame, this object can be used in most sparklyr functions, including data analysis with dplyr or machine learning:

```
library(sparklyr)
sc <- spark_connect(master = "local")

iris_tbl <- copy_to(sc, iris, overwrite = TRUE)
```

You can inspect which tables are cached by navigating to the Spark UI using spark_web(sc), clicking the Storage tab, and then clicking on a specific RDD, as illustrated in Figure 9-7.

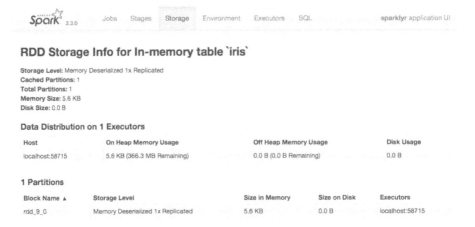

RDD Storage Info for In-memory table `iris`

Storage Level: Memory Deserialized 1x Replicated
Cached Partitions: 1
Total Partitions: 1
Memory Size: 5.6 KB
Disk Size: 0.0 B

Data Distribution on 1 Executors

Host	On Heap Memory Usage	Off Heap Memory Usage	Disk Usage
localhost:58715	5.6 KB (366.3 MB Remaining)	0.0 B (0.0 B Remaining)	0.0 B

1 Partitions

Block Name ▲	Storage Level	Size in Memory	Size on Disk	Executors
rdd_9_0	Memory Deserialized 1x Replicated	5.6 KB	0.0 B	localhost:58715

Figure 9-7. Cached RDD in the Spark web interface

Data loaded in memory will be released when the R session terminates, either explicitly or implicitly, with a restart or disconnection; however, to free up resources, you can use `tbl_uncache()`:

```
tbl_uncache(sc, "iris")
```

Checkpointing

Checkpointing is a slightly different type of caching; while it also saves data, it will additionally break the graph computation lineage. For example, if a cached partition is lost, it can be computed from the computation graph, which is not possible with checkpointing since the source of computation is lost.

When performing operations which create expensive computation graphs, it can make sense to checkpoint to save and break the computation lineage in order to help Spark reduce graph computation resources; otherwise, Spark might try to optimize a computation graph that is really not useful to optimize.

You can checkpoint explicitly by saving to CSV, Parquet, and other file formats. Or, let Spark checkpoint this for you by using `sdf_checkpoint()` in `sparklyr`, as follows:

```
# set checkpoint path
spark_set_checkpoint_dir(sc, getwd())

# checkpoint the iris dataset
iris_tbl %>% sdf_checkpoint()
```

Notice that checkpointing truncates the computation lineage graph, which can speed up performance if the same intermediate result is used multiple times.

Memory

Memory in Spark is categorized into *reserved*, *user*, *execution*, or *storage*:

Reserved
Reserved memory is the memory Spark needs to function and therefore is overhead that is required and should not be configured. This value defaults to 300 MB.

User
User memory is the memory used to execute custom code. `sparklyr` makes use of this memory only indirectly when executing `dplyr` expressions or modeling a dataset.

Execution
Execution memory is used to execute code by Spark, mostly to process the results from the partition and perform shuffling.

Storage
Storage memory is used to cache RDDs—for instance, when using `compute()` in `sparklyr`.

As part of tuning execution, you can consider tweaking the amount of memory allocated for user, execution, and storage by creating a Spark connection with different values than the defaults provided in Spark:

```
config <- spark_config()

# define memory available for storage and execution
config$spark.memory.fraction <- 0.75

# define memory available for storage
config$spark.memory.storageFraction <- 0.5
```

For instance, if you want to use Spark to store large amounts of data in memory with the purpose of quickly filtering and retrieving subsets, you can expect Spark to use little execution or user memory. Therefore, to maximize storage memory, you can tune Spark as follows:

```
config <- spark_config()

# define memory available for storage and execution
config$spark.memory.fraction <- 0.90

# define memory available for storage
config$spark.memory.storageFraction <- 0.90
```

However, note that Spark will borrow execution memory from storage and vice versa if needed and if possible; therefore, in practice, there should be little need to tune the memory settings.

Shuffling

Shuffling is the operation that redistributes data across machines; it is usually expensive and therefore something you should try to minimize. You can easily identify whether significant time is being spent shuffling by looking at the event timeline. It is possible to reduce shuffling by reframing data analysis questions or hinting Spark appropriately.

This would be relevant, for instance, when joining DataFrames that differ in size significantly; that is, one set is orders of magnitude smaller than the other one. You can consider using sdf_broadcast() to mark a DataFrame as small enough for use in broadcast joins, meaning it pushes one of the smaller DataFrames to each of the worker nodes to reduce shuffling the bigger DataFrame. Here's one example for sdf_broadcast():

```
sdf_len(sc, 10000) %>%
    sdf_broadcast() %>%
    left_join(sdf_len(sc, 100))
```

Serialization

Serialization is the process of translating data and tasks into a format that can be transmitted between machines and reconstructed on the receiving end.

It is not that common to need to adjust serialization when tuning Spark; however, it is worth mentioning that there are alternative serialization modules like the Kryo Serializer (*https://oreil.ly/TRbNh*) that can provide performance improvements over the default Java Serializer (*https://oreil.ly/0DMsd*).

You can turn on the Kryo Serializer in sparklyr through the following:

```
config <- spark_config()

config$spark.serializer <- "org.apache.spark.serializer.KryoSerializer"
sc <- spark_connect(master = "local", config = config)
```

Configuration Files

Configuring the spark_config() settings before connecting is the most common approach while tuning Spark. However, after you identify the parameters in your connection, you should consider switching to use a configuration file since it will remove the clutter in your connection code and also allow you to share the configuration settings across projects and coworkers.

For instance, instead of connecting to Spark like this:

```
config <- spark_config()
config["sparklyr.shell.driver-memory"] <- "2G"
sc <- spark_connect(master = "local", config = config)
```

you can define a *config.yml* file with the desired settings. This file should be located in the current working directory or in parent directories. For example, you can create the following *config.yml* file to modify the default driver memory:

```
default:
  sparklyr.shell.driver-memory: 2G
```

Then, connecting with the same configuration settings becomes much cleaner by using instead:

```
sc <- spark_connect(master = "local")
```

You can also specify an alternate configuration filename or location by setting the file parameter in spark_config(). One additional benefit from using configuration files is that a system administrator can change the default configuration by changing the value of the R_CONFIG_ACTIVE environment variable. See the GitHub rstudio/config (*https://oreil.ly/74jIL*) repo for additional information.

Recap

This chapter provided a broad overview of Spark internals and detailed configuration settings to help you speed up computation and enable high computation loads. It provided the foundations to understand bottlenecks and guidance on common configuration considerations. However, fine-tuning Spark is a broad topic that would require many more chapters to cover extensively. Therefore, while troubleshooting Spark's performance and scalability, searching the web, and consulting online communities, it is often necessary to fine-tune your particular environment as well.

Chapter 10 introduces the ecosystem of Spark extensions that are available in R. Most extensions are highly specialized, but they will prove to be extremely useful in specific cases and for readers with particular needs. For instance, they can process nested data, perform graph analysis, and use different modeling libraries like rsparkling from H20. In addition, the next few chapters introduce many advanced data analysis and modeling topics that are required to master large-scale computing in R.

## CHAPTER 10

Extensions

*I try to know as many people as I can. You never
know which one you'll need.*

—*Tyrion Lannister*

In Chapter 9, you learned how Spark processes data at large scale by allowing users to configure the cluster resources, partition data implicitly or explicitly, execute commands across distributed compute nodes, shuffle data across them when needed, cache data to improve performance, and serialize data efficiently over the network. You also learned how to configure the different Spark settings used while connecting, submitting a job, and running an application, as well as particular settings applicable only to R and R extensions that we present in this chapter.

Chapters 3, 4, and 8 provided a foundation to read and understand most datasets. However, the functionality that was presented was scoped to Spark's built-in features and tabular datasets. This chapter goes beyond tabular data and explores how to analyze and model networks of interconnected objects through graph processing, analyze genomics datasets, prepare data for deep learning, analyze geographic datasets, and use advanced modeling libraries like H2O and XGBoost over large-scale datasets.

The combination of all the content presented in the previous chapters should take care of most of your large-scale computing needs. However, for those few use cases for which functionality is still lacking, the following chapters provide tools to extend Spark yourself—through custom R transformation, custom Scala code, or a recent new execution mode in Spark that enables analyzing real-time datasets. However, before reinventing the wheel, let's examine some of the extensions available in Spark.

Overview

In Chapter 1, we presented the R community as a vibrant group of individuals collaborating with each other in many ways—for example, moving open science forward by creating R packages that you can install from CRAN. In a similar way, but at a much smaller scale, the R community has contributed extensions that increase the functionality initially supported in Spark and R. Spark itself also provides support for creating Spark extensions and, in fact, many R extensions make use of Spark extensions.

Extensions are constantly being created, so this section will be outdated at some given point in time. In addition, we might not even be aware of many Spark and R extensions. However, at the very least we can track the extensions that are available in CRAN by looking at the "reverse imports" for `sparklyr` in CRAN (*http://bit.ly/2Z568xz*). Extensions and R packages published in CRAN tend to be the most stable since when a package is published in CRAN, it goes through a review process that increases the overall quality of a contribution.

While we wish we could present all the extensions, we've instead scoped this chapter to the extensions that should be the most interesting to you. You can find additional extensions at the github.com/r-spark (*https://github.com/r-spark*) organization or by searching repositories on GitHub with the `sparklyr` tag.

rsparkling
> The `rsparkling` extensions allow you to use H2O and Spark from R. This extension is what we would consider advanced modeling in Spark. While Spark's built-in modeling library, Spark MLlib, is quite useful in many cases, H2O's modeling capabilities can compute additional statistical metrics and can provide performance and scalability improvements over Spark MLlib. We have neither performed detailed comparisons nor benchmarks between MLlib and H2O; this is something you will need to research on your own to create a complete picture of when to use H2O's capabilities.

graphframes
> The `graphframes` extensions adds support to process graphs in Spark. A graph is a structure that describes a set of objects in which some pairs of the objects are in some sense related. As you learned in Chapter 1, ranking web pages was an early motivation to develop precursors to Spark powered by MapReduce; web pages happen to form a graph if you consider a link between pages as the relationship between each pair of pages. Computing operations likes PageRank over graphs can be quite useful in web search and social networks, for example.

sparktf
> The `sparktf` extension provides support to write TensorFlow records in Spark. TensorFlow is one of the leading deep learning frameworks, and it is often used

with large amounts of numerical data represented as TensorFlow records, a file format optimized for TensorFlow. Spark is often used to process unstructured and large-scale datasets into smaller numerical datasets that can easily fit into a GPU. You can use this extension to save datasets in the TensorFlow record file format.

xgboost

The xgboost extension brings the well-known XGBoost modeling library to the world of large-scale computing. XGBoost is a scalable, portable, and distributed library for gradient boosting. It became well known in the machine learning competition circles after its use in the winning solution of the Higgs Boson Machine Learning Challenge (*http://bit.ly/2YPE2qO*) and has remained popular in other Kaggle competitions since then.

variantspark

The variantspark extension provides an interface to use Variant Spark, a scalable toolkit for genome-wide association studies (GWAS). It currently provides functionality to build random forest models, estimating variable importance, and reading variant call format (VCF) files. While there are other random forest implementations in Spark, most of them are not optimized to deal with GWAS datasets, which usually come with thousands of samples and millions of variables.

geospark

The geospark extension enables us to load and query large-scale geographic datasets. Usually datasets containing latitude and longitude points or complex areas are defined in the well-known text (WKT) format, a text markup language for representing vector geometry objects on a map.

Before you learn how and when to use each extension, we should first briefly explain how you can use extensions with R and Spark.

First, a Spark extension is just an R package that happens to be aware of Spark. As with any other R package, you will first need to install the extension. After you've installed it, it is important to know that you will need to reconnect to Spark before the extension can be used. So, in general, here's the pattern you should follow:

```
library(sparkextension)
library(sparklyr)

sc <- spark_connect(master = "<master>")
```

Notice that sparklyr is loaded after the extension to allow the extension to register properly. If you had to install and load a new extension, you would first need to disconnect using spark_disconnect(sc), restart your R session, and repeat the preceding steps with the new extension.

It's not difficult to install and use Spark extensions from R; however, each extension can be a world of its own, so most of the time you will need to spend time understanding what the extension is, when to use it, and how to use it properly. The first extension you will learn about is the `rsparkling` extension, which enables you to use H2O in Spark with R.

H2O

H2O (*https://www.h2o.ai/*), created by H2O.ai, is open source software for large-scale modeling that allows you to fit thousands of potential models as part of discovering patterns in data. You can consider using H2O to complement or replace Spark's default modeling algorithms. It is common to use Spark's default modeling algorithms and transition to H2O when Spark's algorithms fall short or when advanced functionality (like additional modeling metrics or automatic model selection) is desired.

We can't do justice to H2O's great modeling capabilities in a single paragraph; explaining H2O properly would require a book in and of itself. Instead, we would like to recommend reading Darren Cook's *Practical Machine Learning with H2O* (*https://oreil.ly/l5RHI*) (O'Reilly) to explore in-depth H2O's modeling algorithms and features. In the meantime, you can use this section as a brief guide to get started using H2O in Spark with R.

To use H2O with Spark, it is important to know that there are four components involved: H2O, Sparkling Water, `rsparkling` (*http://bit.ly/2MlFxqa*), and Spark. Sparkling Water allows users to combine the fast, scalable machine learning algorithms of H2O with the capabilities of Spark. You can think of Sparkling Water as a component bridging Spark with H2O and `rsparkling` as the R frontend for Sparkling Water, as depicted in Figure 10-1.

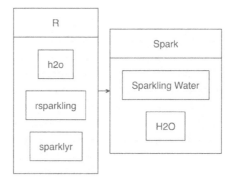

Figure 10-1. H2O components with Spark and R

First, install `rsparkling` and `h2o` as specified on the `rsparkling` documentation site (*http://bit.ly/2Z78MD0*).

```
install.packages("h2o", type = "source",
  repos = "http://h2o-release.s3.amazonaws.com/h2o/rel-yates/5/R")
install.packages("rsparkling", type = "source",
  repos = "http://h2o-release.s3.amazonaws.com/sparkling-water/rel-2.3/31/R")
```

It is important to note that you need to use compatible versions of Spark, Sparkling Water, and H2O as specified in their documentation; we present instructions for Spark 2.3, but using different Spark versions will require you to install different versions. So let's start by checking the version of H2O by running the following:

```
packageVersion("h2o")
```

```
## [1] '3.24.0.5'
```

```
packageVersion("rsparkling")
```

```
## [1] '2.3.31'
```

We then can connect with the supported Spark versions as follows (you will have to adjust the `master` parameter for your particular cluster):

```
library(rsparkling)
library(sparklyr)
library(h2o)

sc <- spark_connect(master = "local", version = "2.3",
                    config = list(sparklyr.connect.timeout = 3 * 60))

cars <- copy_to(sc, mtcars)
```

H2O provides a web interface that can help you monitor training and access much of H2O's functionality. You can access the web interface (called H2O Flow) through `h2o_flow(sc)`, as shown in Figure 10-2.

When using H2O, you will need to convert your Spark DataFrame into an H2O DataFrame through `as_h2o_frame`:

```
cars_h2o <- as_h2o_frame(sc, cars)
cars_h2o

    mpg cyl disp  hp drat    wt  qsec vs am gear carb
1  21.0   6  160 110 3.90 2.620 16.46  0  1    4    4
2  21.0   6  160 110 3.90 2.875 17.02  0  1    4    4
3  22.8   4  108  93 3.85 2.320 18.61  1  1    4    1
4  21.4   6  258 110 3.08 3.215 19.44  1  0    3    1
5  18.7   8  360 175 3.15 3.440 17.02  0  0    3    2
6  18.1   6  225 105 2.76 3.460 20.22  1  0    3    1

[32 rows x 11 columns]
```

Figure 10-2. The H2O Flow interface using Spark with R

Then, you can use many of the modeling functions available in the h2o package with ease. For instance, we can fit a generalized linear model with ease:

```
model <- h2o.glm(x = c("wt", "cyl"),
                 y = "mpg",
                 training_frame = cars_h2o,
                 lambda_search = TRUE)
```

H2O provides additional metrics not necessarily available in Spark's modeling algorithms. The model that we just fit, `Residual Deviance`, is provided in the model, while this would not be a standard metric when using Spark MLlib.

```
model

...
MSE:  6.017684
RMSE:  2.453097
MAE:  1.940985
RMSLE:  0.1114801
Mean Residual Deviance :  6.017684
R^2 :  0.8289895
Null Deviance :1126.047
Null D.o.F. :31
Residual Deviance :192.5659
Residual D.o.F. :29
AIC :156.2425
```

Then, you can run prediction over the generalized linear model (GLM). A similar approach would work for many other models available in H2O:

```
predictions <- as_h2o_frame(sc, copy_to(sc, data.frame(wt = 2, cyl = 6)))
h2o.predict(model, predictions)

   predict
1 24.05984

[1 row x 1 column]
```

You can also use H2O to perform automatic training and tuning of many models, meaning that H2O can choose which model to use for you using AutoML (*https://oreil.ly/Ck9Ao*):

```
automl <- h2o.automl(x = c("wt", "cyl"), y = "mpg",
                     training_frame = cars_h2o,
                     max_models = 20,
                     seed = 1)
```

For this particular dataset, H2O determines that a deep learning model is a better fit than a GLM.[1] Specifically, H2O's AutoML explored using XGBoost, deep learning, GLM, and a Stacked Ensemble model:

```
automl@leaderboard

  model_id              mean_residual_dev…    rmse      mse      mae     rmsle
1 DeepLearning_…                 6.541322 2.557601 6.541322 2.192295 0.1242028
2 XGBoost_grid_1…                6.958945 2.637981 6.958945 2.129421 0.1347795
3 XGBoost_grid_1_…               6.969577 2.639996 6.969577 2.178845 0.1336290
4 XGBoost_grid_1_…               7.266691 2.695680 7.266691 2.167930 0.1331849
5 StackedEnsemble…               7.304556 2.702694 7.304556 1.938982 0.1304792
6 XGBoost_3_…                    7.313948 2.704431 7.313948 2.088791 0.1348819
```

Rather than using the leaderboard, you can focus on the best model through automl@leader; for example, you can glance at the particular parameters from this deep learning model as follows:

```
tibble::tibble(parameter = names(automl@leader@parameters),
               value = as.character(automl@leader@parameters))

# A tibble: 20 x 2
   parameter                          values
   <chr>                              <chr>
 1 model_id                           DeepLearning_grid_1_AutoML…
 2 training_frame                     automl_training_frame_rdd…
 3 nfolds                             5
 4 keep_cross_validation_models       FALSE
 5 keep_cross_validation_predictions  TRUE
 6 fold_assignment                    Modulo
 7 overwrite_with_best_model          FALSE
 8 activation                         RectifierWithDropout
 9 hidden                             200
```

---

1 Notice that AutoML uses cross-validation, which we did not use in GLM.

```
10 epochs                   10003.6618461538
11 seed                     1
12 rho                      0.95
13 epsilon                  1e-06
14 input_dropout_ratio      0.2
15 hidden_dropout_ratios    0.4
16 stopping_rounds          0
17 stopping_metric          deviance
18 stopping_tolerance       0.05
19 x                        c("cyl", "wt")
20 y                        mpg
```

You can then predict using the leader as follows:

```
h2o.predict(automl@leader, predictions)

  predict
1 30.74639

[1 row x 1 column]
```

Many additional examples are available (*http://bit.ly/2NdTIwX*), and you can also request help from the official GitHub repository (*http://bit.ly/2MlFxqa*) for the `rspar kling` package.

The next extension, `graphframes`, allows you to process large-scale relational data-sets. Before you start using it, make sure to disconnect with `spark_disconnect(sc)` and restart your R session since using a different extension requires you to reconnect to Spark and reload `sparklyr`.

Graphs

The first paper in the history of graph theory was written by Leonhard Euler on the Seven Bridges of Königsberg in 1736. The problem was to devise a walk through the city that would cross each bridge once and only once. Figure 10-3 presents the original diagram.

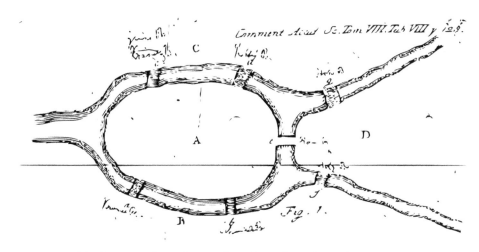

Figure 10-3. *The Seven Bridges of Königsberg from the Euler archive*

Today, a graph is defined as an ordered pair $G = (V, E)$, with V a set of vertices (nodes or points) and $E \subseteq \{\{x, y\} \mid (x, y) \in V^2 \land x \neq y\}$ a set of edges (links or lines), which are either an unordered pair for *undirected graphs* or an ordered pair for *directed graphs*. The former describes links where the direction does not matter, and the latter describes links where it does.

As a simple example, we can use the `highschool` dataset from the `ggraph` package, which tracks friendship among high school boys. In this dataset, the vertices are the students and the edges describe pairs of students who happen to be friends in a particular year:

```
install.packages("ggraph")
install.packages("igraph")

ggraph::highschool
```

```
# A tibble: 506 x 3
    from     to year
   <dbl>  <dbl> <dbl>
 1     1     14  1957
 2     1     15  1957
 3     1     21  1957
 4     1     54  1957
 5     1     55  1957
 6     2     21  1957
 7     2     22  1957
 8     3      9  1957
 9     3     15  1957
10     4      5  1957
# … with 496 more rows
```

While the high school dataset can easily be processed in R, even medium-size graph datasets can be difficult to process without distributing this work across a cluster of machines, for which Spark is well suited. Spark supports processing graphs through the graphframes (*http://bit.ly/2Z5hVYB*) extension, which in turn uses the GraphX (*http://bit.ly/30cbKU6*) Spark component. GraphX is Apache Spark's API for graphs and graph-parallel computation. It's comparable in performance to the fastest specialized graph-processing systems and provides a growing library of graph algorithms.

A graph in Spark is also represented as a DataFrame of edges and vertices; however, our format is slightly different since we will need to construct a DataFrame for vertices. Let's first install the graphframes (*http://bit.ly/2Z5hVYB*) extension:

```
install.packages("graphframes")
```

Next, we need to connect, copying the highschool dataset and transforming the graph to the format that this extension expects. Here, we scope this dataset to the friendships of the year 1957:

```
library(graphframes)
library(sparklyr)
library(dplyr)

sc <- spark_connect(master = "local", version = "2.3")
highschool_tbl <- copy_to(sc, ggraph::highschool, "highschool") %>%
  filter(year == 1957) %>%
  transmute(from = as.character(as.integer(from)),
            to = as.character(as.integer(to)))

from_tbl <- highschool_tbl %>% distinct(from) %>% transmute(id = from)
to_tbl <- highschool_tbl %>% distinct(to) %>% transmute(id = to)

vertices_tbl <- distinct(sdf_bind_rows(from_tbl, to_tbl))
edges_tbl <- highschool_tbl %>% transmute(src = from, dst = to)
```

The vertices_tbl table is expected to have a single id column:

```
vertices_tbl

# Source: spark<?> [?? x 1]
   id
   <chr>
 1 1
 2 34
 3 37
 4 43
 5 44
 6 45
 7 56
 8 57
 9 65
10 71
# … with more rows
```

And the `edges_tbl` is expected to have `src` and `dst` columns:

```
edges_tbl

# Source: spark<?> [?? x 2]
    src   dst
    <chr> <chr>
 1  1     14
 2  1     15
 3  1     21
 4  1     54
 5  1     55
 6  2     21
 7  2     22
 8  3     9
 9  3     15
10  4     5
# … with more rows
```

You can now create a GraphFrame:

```
graph <- gf_graphframe(vertices_tbl, edges_tbl)
```

We now can use this graph to start analyzing this dataset. For instance, we'll find out how many friends on average every boy has, which is referred to as the *degree* or *valency* of a *vertex*:

```
gf_degrees(graph) %>% summarise(friends = mean(degree))

# Source: spark<?> [?? x 1]
  friends
    <dbl>
1    6.94
```

We then can find what the shortest path to some specific vertex (a boy for this dataset). Since the data is anonymized, we can just pick the boy identified as 33 and find how many degrees of separation exist between them:

```
gf_shortest_paths(graph, 33) %>%
  filter(size(distances) > 0) %>%
  mutate(distance = explode(map_values(distances))) %>%
  select(id, distance)

# Source: spark<?> [?? x 2]
   id    distance
   <chr>    <int>
 1 19          5
 2 5           4
 3 27          6
 4 4           4
 5 11          6
 6 23          4
 7 36          1
 8 26          2
```

```
 9 33          0
10 18          5
# … with more rows
```

Finally, we can also compute PageRank over this graph, which was presented in Chapter 1's discussion of Google's page ranking algorithm:

```
gf_graphframe(vertices_tbl, edges_tbl) %>%
  gf_pagerank(reset_prob = 0.15, max_iter = 10L)

GraphFrame
Vertices:
  Database: spark_connection
  $ id       <dbl> 12, 12, 14, 14, 27, 27, 55, 55, 64, 64, 41, 41, 47, 47, 6…
  $ pagerank <dbl> 0.3573460, 0.3573460, 0.3893665, 0.3893665, 0.2362396, 0.…
Edges:
  Database: spark_connection
  $ src    <dbl> 7, 7, 7, 7, 7, 7, 7, 7, 7, 7, 7, 7, 7, 7, 7, 7, 12, 12, 12,…
  $ dst    <dbl> 17, 17, 17, 17, 17, 17, 17, 17, 17, 17, 17, 17, 17, 17, 17,…
  $ weight <dbl> 0.25000000, 0.25000000, 0.25000000, 0.25000000, 0.25000000,…
```

To give you some insights into this dataset, Figure 10-4 plots this chart using the ggraph and highlights the highest PageRank scores for the following dataset:

```
highschool_tbl %>%
  igraph::graph_from_data_frame(directed = FALSE) %>%
  ggraph(layout = 'kk') +
    geom_edge_link(alpha = 0.2,
                   arrow = arrow(length = unit(2, 'mm')),
                   end_cap = circle(2, 'mm'),
                   start_cap = circle(2, 'mm')) +
    geom_node_point(size = 2, alpha = 0.4)
```

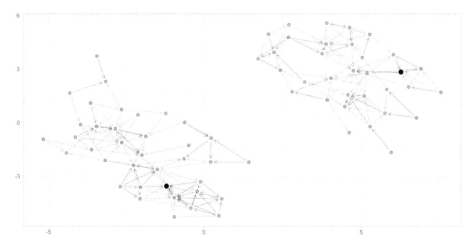

Figure 10-4. High school ggraph dataset with highest PageRank highlighted

There are many more graph algorithms provided in graphframes—for example, breadth-first search, connected components, label propagation for detecting communities, strongly connected components, and triangle count. For questions on this extension refer to the official GitHub repository (*http://bit.ly/2Z5hVYB*). We now present a popular gradient-boosting framework—make sure to disconnect and restart before trying the next extension.

XGBoost

A *decision tree* is a flowchart-like structure in which each internal node represents a test on an attribute, each branch represents the outcome of the test, and each leaf node represents a class label. For example, Figure 10-5 shows a decision tree that could help classify whether an employee is likely to leave given a set of factors like job satisfaction and overtime. When a decision tree is used to predict continuous variables instead of discrete outcomes—say, how likely someone is to leave a company—it is referred to as a *regression tree*.

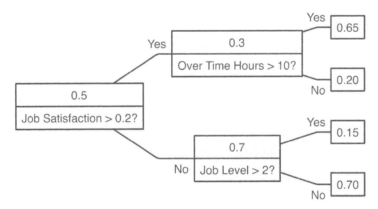

Figure 10-5. A decision tree to predict job attrition based on known factors

While a decision tree representation is quite easy to understand and to interpret, finding out the decisions in the tree requires mathematical techniques like *gradient descent* to find a local minimum. Gradient descent takes steps proportional to the negative of the gradient of the function at the current point. The gradient is represented by ∇, and the learning rate by γ. You simply start from a given state a_n and compute the next iteration a_{n+1} by following the direction of the gradient:

$$a_{n+1} = a_n - \gamma \nabla F(a_n)$$

XGBoost is an open source software library that provides a gradient-boosting framework. It aims to provide scalable, portable, and distributed gradient boosting for training gradient-boosted decision trees (GBDT) and gradient-boosted regression trees (GBRT). Gradient-boosted means XGBoost uses gradient descent and boosting, which is a technique that chooses each predictor sequentially.

sparkxgb is an extension that you can use to train XGBoost models in Spark; however, be aware that currently Windows is unsupported. To use this extension, first install it from CRAN:

```
install.packages("sparkxgb")
```

Then, you need to import the sparkxgb extension followed by your usual Spark connection code, adjusting master as needed:

```
library(sparkxgb)
library(sparklyr)
library(dplyr)

sc <- spark_connect(master = "local", version = "2.3")
```

For this example, we use the attrition dataset from the rsample package, which you would need to install by using install.packages("rsample"). This is a fictional dataset created by IBM data scientists to uncover the factors that lead to employee attrition:

```
attrition <- copy_to(sc, rsample::attrition)
attrition

# Source: spark<?> [?? x 31]
     Age Attrition BusinessTravel DailyRate Department DistanceFromHome
   <int> <chr>     <chr>              <int> <chr>                 <int>
1     41 Yes       Travel_Rarely       1102 Sales                    1
2     49 No        Travel_Freque…       279 Research_…               8
3     37 Yes       Travel_Rarely       1373 Research_…               2
4     33 No        Travel_Freque…      1392 Research_…               3
5     27 No        Travel_Rarely        591 Research_…               2
6     32 No        Travel_Freque…      1005 Research_…               2
7     59 No        Travel_Rarely       1324 Research_…               3
8     30 No        Travel_Rarely       1358 Research_…              24
9     38 No        Travel_Freque…       216 Research_…              23
10    36 No        Travel_Rarely       1299 Research_…              27
# … with more rows, and 25 more variables: Education <chr>,
#   EducationField <chr>, EnvironmentSatisfaction <chr>, Gender <chr>,
#   HourlyRate <int>, JobInvolvement <chr>, JobLevel <int>, JobRole <chr>,
#   JobSatisfaction <chr>, MaritalStatus <chr>, MonthlyIncome <int>,
#   MonthlyRate <int>, NumCompaniesWorked <int>, OverTime <chr>,
#   PercentSalaryHike <int>, PerformanceRating <chr>,
#   RelationshipSatisfaction <chr>, StockOptionLevel <int>,
#   TotalWorkingYears <int>, TrainingTimesLastYear <int>,
#   WorkLifeBalance <chr>, YearsAtCompany <int>, YearsInCurrentRole <int>,
#   YearsSinceLastPromotion <int>, YearsWithCurrManager <int>
```

To build an XGBoost model in Spark, use xgboost_classifier(). We will compute attrition against all other features by using the Attrition ~ . formula and specify 2 for the number of classes since the attrition attribute tracks only whether an employee leaves or stays. Then, you can use `ml_predict()` to predict over large-scale datasets:

```
xgb_model <- xgboost_classifier(attrition,
                                Attrition ~ .,
                                num_class = 2,
                                num_round = 50,
                                max_depth = 4)

xgb_model %>%
  ml_predict(attrition) %>%
  select(Attrition, predicted_label, starts_with("probability_")) %>%
  glimpse()

Observations: ??
Variables: 4
Database: spark_connection
$ Attrition       <chr> "Yes", "No", "Yes", "No", "No", "No", "No", "No", "No", …
$ predicted_label <chr> "No", "Yes", "No", "Yes", "Yes", "Yes", "Yes", "Yes", "Y…
$ probability_No  <dbl> 0.753938094, 0.024780750, 0.915146366, 0.143568754, 0.07…
$ probability_Yes <dbl> 0.24606191, 0.97521925, 0.08485363, 0.85643125, 0.927375…
```

XGBoost became well known in the competition circles after its use in the winning solution of the Higgs Machine Learning Challenge, which uses the ATLAS experiment to identify the Higgs boson. Since then, it has become a popular model and used for a large number of Kaggle competitions. However, decision trees could prove limiting especially in datasets with nontabular data like images, audio, and text, which you can better tackle with deep learning models—should we remind you to disconnect and restart?

Deep Learning

A *perceptron* is a mathematical model introduced by Frank Rosenblatt,[2] who developed it as a theory for a hypothetical nervous system. The perceptron maps stimuli to numeric inputs that are weighted into a threshold function that activates only when enough stimuli is present, mathematically:

$$f(x) = \begin{cases} 1 & \sum_{i=1}^{m} w_i x_i + b > 0 \\ 0 & \text{otherwise} \end{cases}$$

Minsky and Papert found out that a single perceptron can classify only datasets that are linearly separable; however, they also revealed in their book *Perceptrons* that

---

2 Rosenblatt F (1958). "The perceptron: a probabilistic model for information storage and organization in the brain." *Psychological review.*

layering perceptrons would bring additional classification capabilities.[3] Figure 10-6 presents the original diagram showcasing a multilayered perceptron.

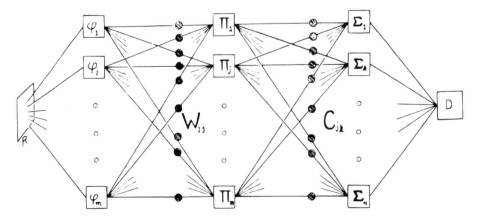

Figure 10-6. Layered perceptrons, as illustrated in the book Perceptrons

Before we start, let's first install all the packages that we are about to use:

```
install.packages("sparktf")
install.packages("tfdatasets")
```

Using Spark we can create a multilayer perceptron classifier with `ml_multilayer_per ceptron_classifier()` and gradient descent to classify and predict over large datasets. Gradient descent was introduced to layered perceptrons by Geoff Hinton.[4]

```
library(sparktf)
library(sparklyr)

sc <- spark_connect(master = "local", version = "2.3")

attrition <- copy_to(sc, rsample::attrition)

nn_model <- ml_multilayer_perceptron_classifier(
  attrition,
  Attrition ~ Age + DailyRate + DistanceFromHome + MonthlyIncome,
  layers = c(4, 3, 2),
  solver = "gd")

nn_model %>%
  ml_predict(attrition) %>%
```

3 Minsky M, Papert SA (2017). *Perceptrons: An introduction to computational geometry.* MIT press.

4 Ackley DH, Hinton GE, Sejnowski TJ (1985). "A learning algorithm for Boltzmann machines." *Cognitive science.*

```
select(Attrition, predicted_label, starts_with("probability_")) %>%
  glimpse()

Observations: ??
Variables: 4
Database: spark_connection
$ Attrition      <chr> "Yes", "No", "Yes", "No", "No", "No", "No", "No", "No"…
$ predicted_label <chr> "No", "No", "No", "No", "No", "No", "No", "No", "No", …
$ probability_No  <dbl> 0.8439275, 0.8439275, 0.8439275, 0.8439275, 0.8439275,…
$ probability_Yes <dbl> 0.1560725, 0.1560725, 0.1560725, 0.1560725, 0.1560725,…
```

Notice that the columns must be numeric, so you will need to manually convert them with the feature transforming techniques presented in Chapter 4. It is natural to try to add more layers to classify more complex datasets; however, adding too many layers will cause the gradient to vanish, and other techniques will need to use these deep layered networks, also known as *deep learning models.*

Deep learning models solve the vanishing gradient problem by making use of special activation functions, dropout, data augmentation and GPUs. You can use Spark to retrieve and preprocess large datasets into numerical-only datasets that can fit in a GPU for deep learning training. TensorFlow is one of the most popular deep learning frameworks. As mentioned previously, it supports a binary format known as TensorFlow records.

You can write TensorFlow records using the `sparktf` in Spark, which you can prepare to process in GPU instances with libraries like Keras or TensorFlow.

You can then preprocess large datasets in Spark and write them as TensorFlow records using `spark_write_tf()`:

```
copy_to(sc, iris) %>%
  ft_string_indexer_model(
    "Species", "label",
    labels = c("setosa", "versicolor", "virginica")
  ) %>%
  spark_write_tfrecord(path = "tfrecord")
```

After you have trained the dataset with Keras or TensorFlow, you can use the `tfdata sets` package to load it. You will also need to install the TensorFlow runtime by using `install_tensorflow()` and install Python on your own. To learn more about training deep learning models with Keras, we recommend reading *Deep Learning with R*.[5]

```
tensorflow::install_tensorflow()
tfdatasets::tfrecord_dataset("tfrecord/part-r-00000")

<DatasetV1Adapter shapes: (), types: tf.string>
```

---

5 Chollet F, Allaire J (2018). *Deep Learning with R*. Manning Publications.

Training deep learning models in a single local node with one or more GPUs is often enough for most applications; however, recent state-of-the-art deep learning models train using distributed computing frameworks like Apache Spark. Distributed computing frameworks are used to achieve higher petaflops each day the systems spends training these models. OpenAI (*http://bit.ly/2HawofQ*) analyzed trends in the field of *artificial intelligence* (AI) and cluster computing (illustrated in Figure 10-7). It should be obvious from the figure that there is a trend in recent years to use distributed computing frameworks.

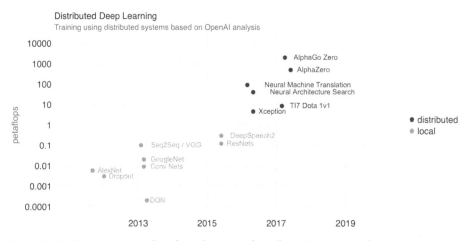

Figure 10-7. Training using distributed systems based on OpenAI analysis

Training large-scale deep learning models is possible in Spark and TensorFlow through frameworks like Horovod. Today, it's possible to use Horovod with Spark from R using the `reticulate` package, since Horovod requires Python and Open MPI, this goes beyond the scope of this book. Next, we will introduce a different Spark extension in the domain of genomics.

Genomics

The human genome (*http://bit.ly/2z2gMqn*) consists of two copies of about 3 billion base pairs of DNA within the 23 chromosome pairs. Figure 10-8 shows the organization of the genome into chromosomes. DNA strands are composed of nucleotides, each composed of one of four nitrogen-containing nucleobases: cytosine (C), guanine (G), adenine (A), or thymine (T). Since the DNA of all humans is nearly identical, we need to store only the differences from the reference genome in the form of a variant call format (VCF) file.

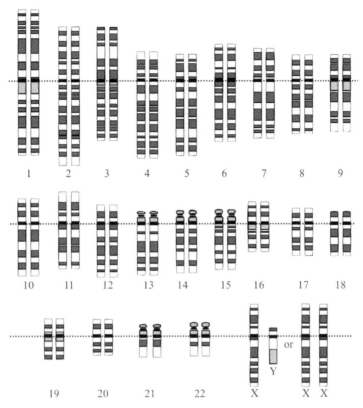

Figure 10-8. The idealized human diploid karyotype showing the organization of the genome into chromosomes

variantspark is a framework based on Scala and Spark to analyze genome datasets. It is being developed by CSIRO Bioinformatics team in Australia. variantspark was tested on datasets with 3,000 samples, each one containing 80 million features in either unsupervised clustering approaches or supervised applications, like classification and regression. variantspark can read VCF files and run analyses while using familiar Spark DataFrames.

To get started, install variantspark from CRAN, connect to Spark, and retrieve a vsc connection to variantspark:

```
library(variantspark)
library(sparklyr)

sc <- spark_connect(master = "local", version = "2.3",
                    config = list(sparklyr.connect.timeout = 3 * 60))

vsc <- vs_connect(sc)
```

We can start by loading a VCF file:

```
vsc_data <- system.file("extdata/", package = "variantspark")

hipster_vcf <- vs_read_vcf(vsc, file.path(vsc_data, "hipster.vcf.bz2"))
hipster_labels <- vs_read_csv(vsc, file.path(vsc_data, "hipster_labels.txt"))
labels <- vs_read_labels(vsc, file.path(vsc_data, "hipster_labels.txt"))
```

variantspark uses random forest to assign an importance score to each tested variant reflecting its association to the interest phenotype. A variant with higher importance score implies it is more strongly associated with the phenotype of interest. You can compute the importance and transform it into a Spark table, as follows:

```
importance_tbl <- vs_importance_analysis(vsc, hipster_vcf,
                                         labels, n_trees = 100) %>%
  importance_tbl()

importance_tbl

# Source: spark<?> [?? x 2]
   variable     importance
   <chr>             <dbl>
 1 2_109511398 0
 2 2_109511454 0
 3 2_109511463 0.00000164
 4 2_109511467 0.00000309
 5 2_109511478 0
 6 2_109511497 0
 7 2_109511525 0
 8 2_109511527 0
 9 2_109511532 0
10 2_109511579 0
# … with more rows
```

You then can use dplyr and ggplot2 to transform the output and visualize it (see Figure 10-9):

```
library(dplyr)
library(ggplot2)

importance_df <- importance_tbl %>%
  arrange(-importance) %>%
  head(20) %>%
  collect()

ggplot(importance_df) +
  aes(x = variable, y = importance) +
  geom_bar(stat = 'identity') +
  scale_x_discrete(limits =
    importance_df[order(importance_df$importance), 1]$variable) +
  coord_flip()
```

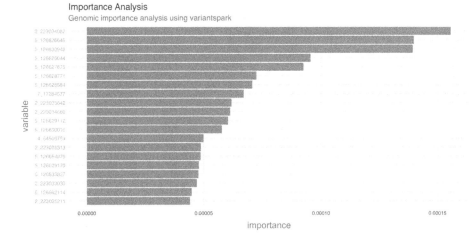

Figure 10-9. Genomic importance analysis using variantspark

This concludes a brief introduction to genomic analysis in Spark using the `variant spark` extension. Next, we move away from microscopic genes to macroscopic datasets that contain geographic locations across the world.

Spatial

`geospark` (*http://bit.ly/2zbTEW8*) enables distributed geospatial computing using a grammar compatible with `dplyr` (*http://bit.ly/2KYKOAC*) and `sf` (*http://bit.ly/2ZerAwb*), which provides a set of tools for working with geospatial vectors.

You can install `geospark` from CRAN, as follows:

```
install.packages("geospark")
```

Then, initialize the `geospark` extension and connect to Spark:

```
library(geospark)
library(sparklyr)

sc <- spark_connect(master = "local", version = "2.3")
```

Next, we load a spatial dataset containing polygons and points:

```
polygons <- system.file("examples/polygons.txt", package="geospark") %>%
  read.table(sep="|", col.names = c("area", "geom"))

points <- system.file("examples/points.txt", package="geospark") %>%
  read.table(sep = "|", col.names = c("city", "state", "geom"))

polygons_wkt <- copy_to(sc, polygons)
points_wkt <- copy_to(sc, points)
```

There are various spatial operations defined in geospark, as depicted in Figure 10-10. These operations allow you to control how geospatial data should be queried based on overlap, intersection, disjoint sets, and so on.

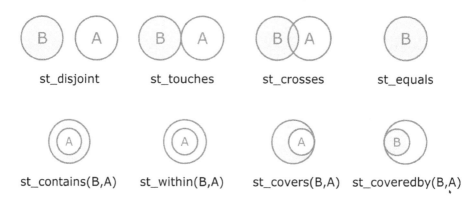

Figure 10-10. Spatial operations available in geospark

For instance, we can use these operations to find the polygons that contain a given set of points using st_contains():

```
library(dplyr)
polygons_wkt <- mutate(polygons_wkt, y = st_geomfromwkt(geom))
points_wkt <- mutate(points_wkt, x = st_geomfromwkt(geom))

inner_join(polygons_wkt,
           points_wkt,
           sql_on = sql("st_contains(y,x)")) %>%
  group_by(area, state) %>%
  summarise(cnt = n())

# Source: spark<?> [?? x 3]
# Groups: area
  area           state   cnt
  <chr>          <chr> <dbl>
1 california area CA       10
2 new york area  NY        9
3 dakota area    ND       10
4 texas area     TX       10
5 dakota area    SD        1
```

You can also plot these datasets by collecting a subset of the entire dataset or aggregating the geometries in Spark before collecting them. One package you should look into is sf.

We close this chapter by presenting a couple of troubleshooting techniques applicable to all extensions.

Troubleshooting

When you are using a new extension for the first time, we recommend increasing the connection timeout (given that Spark will usually need to download extension dependencies) and changing logging to verbose to help you troubleshoot when the download process does not complete:

```
config <- spark_config()
config$sparklyr.connect.timeout <- 3 * 60
config$sparklyr.log.console = TRUE

sc <- spark_connect(master = "local", config = config)
```

In addition, you should know that Apache IVY (*http://ant.apache.org/ivy*) is a popular dependency manager focusing on flexibility and simplicity, and is used by Apache Spark for installing extensions. When the connection fails while you are using an extension, consider clearing your IVY cache (*http://bit.ly/2Zcubun*) by running the following:

```
unlink("~/.ivy2", recursive = TRUE)
```

In addition, you can also consider opening GitHub issues from the following extensions repositories to get help from the extension authors:

- rsparkling (*http://bit.ly/2KUAx8M*)
- sparkxgb (*http://bit.ly/30hG9Ar*)
- sparktf (*http://bit.ly/2z7qNCv*)
- variantspark (*http://bit.ly/2NfjdxX*)
- geospark (*http://bit.ly/2HcgD82*)

Recap

This chapter provided a brief overview on using some of the Spark extensions available in R, which happens to be as easy as installing a package. You then learned how to use the rsparkling extension, which provides access to H2O in Spark, which in turn provides additional modeling functionality like enhanced metrics and the ability to automatically select models. We then jumped to graphframes, an extension to help you process relational datasets that are formally referred to as graphs. You also learned how to compute simple connection metrics or run complex algorithms like PageRank.

The XGBoost and deep learning sections provided alternate modeling techniques that use gradient descent: the former over decision trees, and the latter over deep multi-layered perceptrons where we can use Spark to preprocess datasets into records that later can be consumed by TensorFlow and Keras using the `sparktf` extension. The last two sections introduced extensions to process genomic and spatial datasets through the `variantspark` and `geospark` extensions.

These extensions, and many more, provide a comprehensive library of advanced functionality that, in combination with the analysis and modeling techniques presented, should cover most tasks required to run in computing clusters. However, when functionality is lacking, you can consider writing your own extension, which is what we discuss in Chapter 13, or you can apply custom transformations over each partition using R code, as we describe in Chapter 11.

CHAPTER 11
Distributed R

Not like this. Not like this. Not like this.

—*Cersei Lannister*

In previous chapters, you learned how to perform data analysis and modeling in local Spark instances and proper Spark clusters. In Chapter 10 specifically, we examined how to make use of the additional functionality provided by the Spark and R communities at large. In most cases, the combination of Spark functionality and extensions is more than enough to perform almost any computation. However, for those cases in which functionality is lacking in Spark and their extensions, you could consider distributing R computations to worker nodes yourself.

You can run arbitrary R code in each worker node to run any computation—you can run simulations, crawl content from the web, transform data, and so on. In addition, you can also make use of any package available in CRAN and private packages available in your organization, which reduces the amount of code that you need to write to help you remain productive.

If you are already familiar with R, you might be tempted to use this approach for all Spark operations; however, this is not the recommended use of `spark_apply()`. Previous chapters provided more efficient techniques and tools to solve well-known problems; in contrast, `spark_apply()` introduces additional cognitive overhead, additional troubleshooting steps, performance trade-offs, and, in general, additional complexity you should avoid. Not to say that `spark_apply()` should never be used; rather, `spark_apply()` is reserved to support use cases for which previous tools and techniques fell short.

Overview

Chapter 1 introduced MapReduce as a technique capable of processing large-scale datasets. It also described how Apache Spark provided a superset of operations to perform MapReduce computations easily and more efficiently. Chapter 9 presented insights into how Spark works by applying custom transformations over each partition of the distributed datasets. For instance, if we multiplied each element of a distributed numeric dataset by 10, Spark would apply a mapping operation over each partition through multiple workers. A conceptual view of this process is illustrated in Figure 11-1.

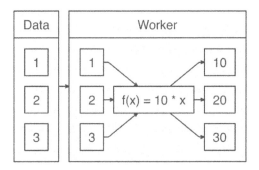

Figure 11-1. Map operation when multiplying by 10

This chapter presents how to define a custom `f(x)` mapping operation using `spark_apply()`; for the previous example, `spark_apply()` provides support to define `10 * x`, as follows:

```
sdf_len(sc, 3) %>% spark_apply(~ 10 * .x)
# Source: spark<?> [?? x 1]
     id
* <dbl>
1    10
2    20
3    30
```

Notice that `~ 10 * .x` is plain R code executed across all worker nodes. The `~` operator is defined in the `rlang` package and provides a compact definition of a function equivalent to `function(.x) 10 * .x`; this compact form is also known as an *anonymous function*, or *lambda expression*.

The `f(x)` function must take an R DataFrame (or something that can be automatically transformed to one) as input and must also produce an R DataFrame as output, as shown in Figure 11-2.

Figure 11-2. Expected function signature in spark_apply() mappings

We can refer back to the original MapReduce example from Chapter 1, where the map operation was defined to split sentences into words and the total unique words were counted as the reduce operation.

In R, we could make use of the unnest_tokens() function from the tidytext R package, which you would need to install from CRAN before connecting to Spark. You can then use tidytext with spark_apply() to tokenize those sentences into a table of words:

```
sentences <- copy_to(sc, data_frame(text = c("I like apples", "I like bananas")))

sentences %>%
  spark_apply(~tidytext::unnest_tokens(.x, word, text))
# Source: spark<?> [?? x 1]
  word
* <chr>
1 i
2 like
3 apples
4 i
5 like
6 bananas
```

We can complete this MapReduce example by performing the reduce operation with dplyr, as follows:

```
sentences %>%
  spark_apply(~tidytext::unnest_tokens(., word, text)) %>%
  group_by(word) %>%
  summarise(count = count())
# Source: spark<?> [?? x 2]
  word     count
* <chr>    <dbl>
1 i            2
2 apples       1
3 like         2
4 bananas      1
```

The rest of this chapter will explain in detail the use cases, features, caveats, considerations, and troubleshooting techniques required when you are defining custom mappings through spark_apply().

The previous sentence tokenizer example can be more efficiently implemented using concepts from previous chapters, specifically through `sentences %>% ft_tokenizer("text", "words") %>% transmute(word = explode(words))`.

Use Cases

Now that we've presented an example to help you understand how `spark_apply()` works, we'll cover a few practical use cases for it:

Import
> You can consider using R to import data from external data sources and formats. For example, when a file format is not natively supported in Spark or its extensions, you can consider using R code to implement a distributed *custom parser* using R packages.

Model
> It is natural to use the rich modeling capabilities already available in R with Spark. In most cases, R models can't be used across large data; however, we will present two particular use cases where R models can be useful at scale. For instance, when data fits into a single machine, you can use *grid search* to optimize their parameters in parallel. In cases where the data can be partitioned to create several models over subsets of the data, you can use *partitioned modeling* in R to compute models across partitions.

Transform
> You can use R's rich data transformation capabilities to complement Spark. We'll present a use case of evaluating data by external systems, and use R to interoperate with them by calling them through a *web API*.

Compute
> When you need to perform large-scale computation in R, or *big compute* as described in Chapter 1, Spark is ideal to distribute this computation. We will present *simulations* as a particular use case for large-scale computing in R.

As we now explore each use case in detail, we'll provide a working example to help you understand how to use `spark_apply()` effectively.

Custom Parsers

Though Spark and its various extensions provide support for many file formats (CSVs, JSON, Parquet, AVRO, etc.), you might need other formats to use at scale. You can parse these additional formats using `spark_apply()` and many existing R packages. In this section, we will look at how to parse logfiles, though similar approaches can be followed to parse other file formats.

It is common to use Spark to analyze logfiles—for instance, logs that track download data from Amazon S3. The `webreadr` package can simplify the process of parsing logs by providing support to load logs stored as Amazon S3, Squid, and the Common log format. You should install `webreadr` from CRAN before connecting to Spark.

For example, an Amazon S3 log looks as follows:

```
#Version: 1.0
#Fields: date time x-edge-location sc-bytes c-ip cs-method cs(Host) cs-uri-stem
  sc-status cs(Referer) cs(User-Agent) cs-uri-query cs(Cookie) x-edge-result-type
  x-edge-request-id x-host-header cs-protocol cs-bytes time-taken

2014-05-23  01:13:11    FRA2    182 192.0.2.10  GET d111111abcdef8.cloudfront.net
  /view/my/file.html    200 www.displaymyfiles.com  Mozilla/4.0%20
  (compatible;%20MSIE%205.0b1;%20Mac_PowerPC)   -   zip=98101   RefreshHit
  MRVMF7KydIvxMWfJIglgwHQwZsbG2IhRJ07sn9AkKUFSHS9EXAMPLE==
  d111111abcdef8.cloudfront.net http   -   0.001
```

This can be parsed easily with `read_aws()`, as follows:

```
aws_log <- system.file("extdata/log.aws", package = "webreadr")
webreadr::read_aws(aws_log)

# A tibble: 2 x 18
  date                edge_location bytes_sent ip_address http_method host  path
  <dttm>              <chr>              <int> <chr>      <chr>       <chr> <chr>
1 2014-05-23 01:13:11 FRA2                 182 192.0.2.10 GET         d111… /vie…
2 2014-05-23 01:13:12 LAX1             2390282 192.0.2.2… GET         d111… /sou…
# … with 11 more variables: status_code <int>, referer <chr>, user_agent <chr>,
#   query <chr>, cookie <chr>, result_type <chr>, request_id <chr>,
#   host_header <chr>, protocol <chr>, bytes_received <chr>, time_elapsed <dbl>
```

To scale this operation, we can make use of `read_aws()` using `spark_apply()`:

```
spark_read_text(sc, "logs", aws_log, overwrite = TRUE, whole = TRUE) %>%
  spark_apply(~webreadr::read_aws(.x$contents))

# Source: spark<?> [?? x 18]
  date                edge_location bytes_sent ip_address http_method host  path
* <dttm>              <chr>              <int> <chr>      <chr>       <chr> <chr>
1 2014-05-23 01:13:11 FRA2                 182 192.0.2.10 GET         d111… /vie…
2 2014-05-23 01:13:12 LAX1             2390282 192.0.2.2… GET         d111… /sou…
# … with 11 more variables: status_code <int>, referer <chr>, user_agent <chr>,
#   query <chr>, cookie <chr>, result_type <chr>, request_id <chr>,
#   host_header <chr>, protocol <chr>, bytes_received <chr>, time_elapsed <dbl>
```

The code used by plain R and `spark_apply()` is similar; however, with `spark_apply()`, logs are parsed in parallel across all the worker nodes available in your cluster.

This concludes the custom parsers discussion; you can parse many other file formats at scale from R following a similar approach. Next we'll look at present partitioned *modeling* as another use case focused on modeling across several datasets in parallel.

Partitioned Modeling

There are many modeling packages available in R that can also be run at scale by partitioning the data into manageable groups that fit in the resources of a single machine.

For instance, suppose that you have a 1 TB dataset for sales data across multiple cities and you are tasked with creating sales predictions over each city. For this case, you can consider partitioning the original dataset per city—say, into 10 GB of data per city —which could be managed by a single compute instance. For this kind of partitionable dataset, you can also consider using spark_apply() by training each model over each city.

As a simple example of partitioned modeling, we can run a linear regression using the iris dataset partitioned by species:

```
iris <- copy_to(sc, datasets::iris)

iris %>%
  spark_apply(nrow, group_by = "Species")
# Source: spark<?> [?? x 2]
  Species      result
  <chr>        <int>
1 versicolor      50
2 virginica       50
3 setosa          50
```

Then you can run a linear regression over each species using spark_apply():

```
iris %>%
  spark_apply(
    function(e) summary(lm(Petal_Length ~ Petal_Width, e))$r.squared,
    names = "r.squared",
    group_by = "Species")
# Source: spark<?> [?? x 2]
  Species      r.squared
  <chr>          <dbl>
1 versicolor     0.619
2 virginica      0.104
3 setosa         0.110
```

As you can see from the r.squared results and intuitively in Figure 11-3, the linear model for versicolor better fits to the regression line:

```
purrr::map(c("versicolor", "virginica", "setosa"),
  ~dplyr::filter(datasets::iris, Species == !!.x) %>%
    ggplot2::ggplot(ggplot2::aes(x = Petal.Length, y = Petal.Width)) +
    ggplot2::geom_point())
```

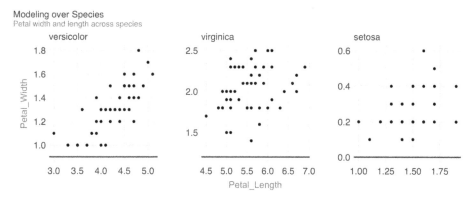

Figure 11-3. Modeling over species in the iris dataset

This concludes our brief overview on how to perform modeling over several different partitionable datasets. A similar technique can be applied to perform modeling over the same dataset using different modeling parameters, as we cover next.

Grid Search

Many R packages provide models that require defining multiple parameters to configure and optimize. When the value of these parameters is unknown, we can distribute this list of unknown parameters across a cluster of machines to find the optimal parameter combination. If the list contains more than one parameter to optimize, it is common to test against all the combinations between parameter A and parameter B, creating a grid of parameters. The process of searching for the best parameter over this parameter grid is commonly known as *grid search*.

For example, we can define a grid of parameters to optimize decision tree models as follows:

```
grid <- list(minsplit = c(2, 5, 10), maxdepth = c(1, 3, 8)) %>%
  purrr:::cross_df() %>%
  copy_to(sc, ., repartition = 9)
grid

# Source: spark<?> [?? x 2]
  minsplit maxdepth
     <dbl>    <dbl>
1        2        1
2        5        1
3       10        1
4        2        3
5        5        3
6       10        3
7        2        8
8        5        8
9       10        8
```

The grid dataset was copied by using `repartition = 9` to ensure that each partition is contained in one machine, since the grid also has nine rows. Now, assuming that the original dataset fits in every machine, we can distribute this dataset to many machines and perform parameter search to find the model that best fits this data:

```
spark_apply(
  grid,
  function(grid, cars) {
    model <- rpart::rpart(
      am ~ hp + mpg,
      data = cars,
      control = rpart::rpart.control(minsplit = grid$minsplit,
                                     maxdepth = grid$maxdepth)
    )
    dplyr::mutate(
      grid,
      accuracy = mean(round(predict(model, dplyr::select(cars, -am))) == cars$am)
    )
  },
  context = mtcars)

# Source: spark<?> [?? x 3]
  minsplit maxdepth accuracy
     <dbl>    <dbl>    <dbl>
1        2        1    0.812
2        5        1    0.812
3       10        1    0.812
4        2        3    0.938
5        5        3    0.938
6       10        3    0.812
7        2        8    1
8        5        8    0.938
9       10        8    0.812
```

For this model, `minsplit = 2` and `maxdepth = 8` produces the most accurate results. You can now use this specific parameter combination to properly train your model.

Web APIs

A web API is a program that can do something useful through a web interface that other programs can reuse. For instance, services like Twitter provide web APIs that allow you to automate reading tweets from a program written in R and other programming languages. You can make use of web APIs using `spark_apply()` by sending programmatic requests to external services using R code.

For example, Google provides a web API to label images using deep learning techniques; you can use this API from R, but for larger datasets, you need to access its APIs from Spark. You can use Spark to prepare data to be consumed by a web API and then use `spark_apply()` to perform this call and process all the incoming results back in Spark.

The next example makes use of the googleAuthR package to authenticate to Google Cloud, the RoogleVision package to perform labeling over the Google Vision API, and spark_apply() to interoperate between Spark and Google's deep learning service. To run the following example, you'll first need to disconnect from Spark and download your *cloudml.json* file from the Google developer portal:

```
sc <- spark_connect(
  master = "local",
  config = list(sparklyr.shell.files = "cloudml.json"))

images <- copy_to(sc, data.frame(
  image = "http://pbs.twimg.com/media/DwzcM88XgAINkg-.jpg"
))

spark_apply(images, function(df) {
  googleAuthR::gar_auth_service(
    scope = "https://www.googleapis.com/auth/cloud-platform",
    json_file = "cloudml.json")

  RoogleVision::getGoogleVisionResponse(
    df$image,
    download = FALSE)
})
# Source: spark<?> [?? x 4]
  mid        description score topicality
  <chr>      <chr>       <dbl>      <dbl>
1 /m/04rky   Mammal      0.973      0.973
2 /m/0bt9lr  Dog         0.958      0.958
3 /m/01z5f   Canidae     0.956      0.956
4 /m/0kpmf   Dog breed   0.909      0.909
5 /m/05mqq3  Snout       0.891      0.891
```

To successfully run a large distributed computation over a web API, the API needs to be able to scale to support the load from all the Spark executors. We can trust that major service providers are likely to support all the requests incoming from your cluster. But when you're calling internal web APIs, make sure the API can handle the load. Also, when you're using third-party services, consider the cost of calling their APIs across all the executors in your cluster to avoid potentially expensive and unexpected charges.

Next we'll describe a use case for big compute where R is used to perform distributed rendering.

Simulations

You can use R combined with Spark to perform large-scale computing. The use case we explore here is rendering computationally expensive images using the rayrender package, which uses *ray tracing*, a photorealistic technique commonly used in movie production.

Let's use this package to render a simple scene that includes a few spheres (see Figure 11-4) that use a *lambertian material*, a diffusely reflecting material or "matte." First, install `rayrender` using `install.packages("rayrender")`. Then, be sure you've disconnected and reconnected Spark:

```
library(rayrender)

scene <- generate_ground(material = lambertian()) %>%
  add_object(sphere(material = metal(color="orange"), z = -2)) %>%
  add_object(sphere(material = metal(color="orange"), z = +2)) %>%
  add_object(sphere(material = metal(color="orange"), x = -2))

render_scene(scene, lookfrom = c(10, 5, 0), parallel = TRUE)
```

Figure 11-4. Ray tracing in Spark using R and rayrender

In higher resolutions, say 1920 x 1080, the previous example takes several minutes to render the single frame from Figure 11-4; rendering a few seconds at 30 frames per second would take several hours in a single machine. However, we can reduce this time using multiple machines by parallelizing computation across them. For instance, using 10 machines with the same number of CPUs would cut rendering time tenfold:

```
system2("hadoop", args = c("fs", "-mkdir", "/rendering"))

sdf_len(sc, 628, repartition = 628) %>%
  spark_apply(function(idx, scene) {
    render <- sprintf("%04d.png", idx$id)
```

```
    rayrender::render_scene(scene, width = 1920, height = 1080,
                            lookfrom = c(12 * sin(idx$id/100),
                                         5, 12 * cos(idx$id/100)),
                            filename = render)

    system2("hadoop", args = c("fs", "-put", render, "/user/hadoop/rendering/"))
}, context = scene, columns = list()) %>% collect()
```

After all the images are rendered, the last step is to collect them from HDFS and use tools like `ffmpeg` to convert individual images into an animation:

```
hadoop fs -get rendering/
ffmpeg -s 1920x1080 -i rendering/%d.png -vcodec libx264 -crf 25
       -pix_fmt yuv420p rendering.mp4
```

 This example assumes HDFS is used as the storage technology for Spark and being run under a `hadoop` user, you will need to adjust this for your particular storage or user.

We've covered some common use cases for `spark_apply()`, but you are certainly welcome to find other use cases for your particular needs. The next sections present technical concepts you'll need to understand to create additional use cases and to use `spark_apply()` effectively.

Partitions

Most Spark operations that analyze data with `dplyr` or model with MLlib don't require understanding how Spark partitions data; they simply work automatically. However, for distributed R computations, this is not the case. For these you will have to learn and understand how exactly Spark is partitioning your data and provide transformations that are compatible with them. This is required since `spark_apply()` receives each partition and allows you to perform any transformation, not the entire dataset. You can refresh concepts like partitioning and transformations using the diagrams and examples from Chapter 9.

To help you understand how partitions are represented in `spark_apply()`, consider the following code:

```
sdf_len(sc, 10) %>%
  spark_apply(~nrow(.x))

# Source: spark<?> [?? x 1]
  result
*  <int>
1      5
2      5
```

Should we expect the output to be the total number of rows? As you can see from the results, in general the answer is no; Spark assumes data will be distributed across multiple machines, so you'll often find it already partitioned, even for small datasets. Because we should not expect `spark_apply()` to operate over a single partition, let's find out how many partitions `sdf_len(sc, 10)` contains:

```
sdf_len(sc, 10) %>% sdf_num_partitions()
```

```
[1] 2
```

This explains why counting rows through `nrow()` under `spark_apply()` retrieves two rows since there are two partitions, not one. `spark_apply()` is retrieving the count of rows over each partition, and each partition contains 5 rows, not 10 rows total, as you might have expected.

For this particular example, we could further aggregate these partitions by repartitioning and then adding up—this would resemble a simple MapReduce operation using `spark_apply()`:

```
sdf_len(sc, 10) %>%
  spark_apply(~nrow(.x)) %>%
  sdf_repartition(1) %>%
  spark_apply(~sum(.x))
```

```
# Source: spark<?> [?? x 1]
  result
*  <int>
1     10
```

So now that you know about partitions using `spark_apply()`, we'll move on to using `group_by` to control partitions.

Grouping

When using `spark_apply()`, we can request explicit partitions from Spark. For instance, if we had to process numbers less than four in one partition and the remaining ones in a second partition, we could create these groups explicitly and then request `spark_apply()` to use them:

```
sdf_len(sc, 10) %>%
  transmute(groups = id < 4) %>%
  spark_apply(~nrow(.x), group_by = "groups")
```

```
# Source: spark<?> [?? x 2]
  groups result
* <lgl>   <int>
1 TRUE        3
2 FALSE       7
```

Notice that `spark_apply()` is still processing two partitions, but in this case we expect these partitions since we explicitly requested them in `spark_apply()`; therefore, you can safely interpret the results as "there are three integers less than four."

 You can only group data by partitions that fit in a single machine; if one of the groups is too large, an exception will be thrown. To perform operations over groups that exceed the resources of a single node, you can consider partitioning to smaller units or use `dplyr::do`, which is currently optimized for large partitions.

The takeaway from this section is to always consider partitions when dealing with `spark_apply()`. Next, we will zoom in to `spark_apply()` to understand how columns are interpreted.

Columns

By default, `spark_apply()` automatically inspects the DataFrame being produced to learn column names and types. For example:

```
sdf_len(sc, 1) %>%
  spark_apply(~ data.frame(numbers = 1, names = "abc"))

# Source: spark<?> [?? x 2]
  numbers names
*   <dbl> <chr>
1       1 abc
```

However, this is inefficient since `spark_apply()` needs to run twice: first, to find columns by computing `spark_apply()` against a subset of all the data, and then to compute the actual desired values.

To improve performance, you can explicitly specify the columns through the `columns` parameters. This parameter takes a named list of types expected in the resulting Data-Frame. We can then rewrite the previous example to run only once by specifying the correct type for the `numbers` column:

```
sdf_len(sc, 1) %>%
  spark_apply(
    ~ data.frame(numbers = 1, names = "abc"),
    columns = list(numbers = "double", names = "character"))

# Source: spark<?> [?? x 2]
  numbers names
*   <dbl> <chr>
1       1 abc
```

Now that we've presented how rows and columns interact with `spark_apply()`, let's move on to making use of the contextual information sometimes required when processing distributed datasets.

Context

To process partitions using `spark_apply()`, you might need to include auxiliary data that is small enough to fit in each node. This was the case in the grid search use case, where the dataset was passed to all partitions and remained unpartitioned itself.

We can modify the initial `f(x) = 10 * x` example in this chapter to customize the multiplier. It was originally set to `10`, but we can make it configurable by specifying it as the `context` parameter:

```
sdf_len(sc, 4) %>%
  spark_apply(
    function(data, context) context * data,
    context = 100
  )
# Source: spark<?> [?? x 1]
      id
   <dbl>
1    100
2    200
3    300
4    400
```

Figure 11-5 illustrates this example conceptually. Notice that the data partitions are still variable; however, the contextual parameter is distributed to all the nodes.

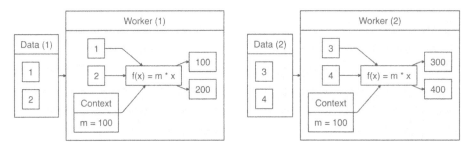

Figure 11-5. Map operation when multiplying with context

The grid search example used this parameter to pass a DataFrame to each worker node; however, since the context parameter is serialized as an R object, it can contain anything. For instance, if you need to pass multiple values—or even multiple datasets—you can pass a list with values.

The following example defines a f(x) = m * x + b function and runs m = 10 and b = 2:

```
sdf_len(sc, 4) %>%
  spark_apply(
    ~.y$m * .x + .y$b,
    context = list(b = 2, m = 10)
  )
# Source: spark<?> [?? x 1]
      id
   <dbl>
1     12
2     22
3     32
4     42
```

Notice that we've renamed context to .y to shorten the variable name. This works because spark_apply() assumes context is the second parameter in functions and expressions.

You'll find the context parameter extremely useful; for instance, the next section presents how to properly construct functions, and context is used in advanced use cases to construct functions dependent on other functions.

Functions

Earlier we presented spark_apply() as an operation to perform custom transformations using a function or expression. In programming literature, functions with a context are also referred to as a *closure*.

Expressions are useful to define short transformations, like ~ 10 * .x. For an expression, .x contains a partition and .y the context, when available. However, it can be hard to define an expression for complex code that spans multiple lines. For those cases, functions are more appropriate.

Functions enable complex and multiline transformations, and are defined as func tion(data, context) {} where you can provide arbitrary code within the {}. We've seen them in previous sections when using Google Cloud to transform images into image captions.

The function passed to spark_apply() is serialized using serialize(), which is described as "a simple low-level interface for serializing to connections." One current limitation of serialize() is that it won't serialize objects being referenced outside its environment. For instance, the following function errors out since the closure references external_value:

```
external_value <- 1
spark_apply(iris, function(e) e + external_value)
```

As workarounds to this limitation, you can add the functions your closure needs into the context and then assign the functions into the global environment:

```
func_a <- function() 40
func_b <- function() func_a() + 1
func_c <- function() func_b() + 1

sdf_len(sc, 1) %>% spark_apply(function(df, context) {
  for (name in names(context)) assign(name, context[[name]], envir = .GlobalEnv)
  func_c()
}, context = list(
  func_a = func_a,
  func_b = func_b,
  func_c = func_c
))
# Source: spark<?> [?? x 1]
  result
  <dbl>
1    42
```

When this isn't feasible, you can also create your own R package with the functionality you need and then use your package in `spark_apply()`.

You've learned all the functionality available in `spark_apply()` using plain R code. In the next section we present how to use packages when distributing computations. R packages are essential when you are creating useful transformations.

Packages

With `spark_apply()` you can use any R package inside Spark. For instance, you can use the `broom` package to create a tidy DataFrame from linear regression output:

```
spark_apply(
  iris,
  function(e) broom::tidy(lm(Petal_Length ~ Petal_Width, e)),
  names = c("term", "estimate", "std.error", "statistic", "p.value"),
  group_by = "Species")
# Source: spark<?> [?? x 6]
  Species    term        estimate std.error statistic  p.value
  <chr>      <chr>          <dbl>     <dbl>     <dbl>    <dbl>
1 versicolor (Intercept)    1.78     0.284      6.28 9.48e- 8
2 versicolor Petal_Width    1.87     0.212      8.83 1.27e-11
3 virginica  (Intercept)    4.24     0.561      7.56 1.04e- 9
4 virginica  Petal_Width    0.647    0.275      2.36 2.25e- 2
5 setosa     (Intercept)    1.33     0.0600    22.1  7.68e-27
6 setosa     Petal_Width    0.546    0.224      2.44 1.86e- 2
```

The first time you call `spark_apply()`, all the contents in your local `.libPaths()` (which contains all R packages) will be copied into each Spark worker node. Packages are only copied once and persist as long as the connection remains open. It's not uncommon for R libraries to be several gigabytes in size, so be prepared for a one-time tax while the R packages are copied over to your Spark cluster. You can disable package distribution by setting `packages = FALSE`.

 Since packages are copied only once for the duration of the `spark_connect()` connection, installing additional packages is not supported while the connection is active. Therefore, if a new package needs to be installed, `spark_disconnect()` the connection, modify packages, and then reconnect. In addition, R packages are not copied in local mode, because the packages already exist on the local system.

Though this section was brief, using packages with distributed R code opens up an entire new universe of interesting use cases. Some of those use cases were presented in this chapter, but by looking at the rich ecosystem of R packages available today you'll find many more.

This section completes our discussion of the functionality needed to distribute R code with R packages. We'll now cover some of the requirements your cluster needs to make use of `spark_apply()`.

Cluster Requirements

The functionality presented in previous chapters did not require special configuration of your Spark cluster—as long as you had a properly configured Spark cluster, you could use R with it. However, for the functionality presented here, your cluster administrator, cloud provider, or you will have to configure your cluster by installing either:

- R in every node, to execute R code across your cluster
- Apache Arrow in every node when using Spark 2.3 or later (Arrow provides performance improvements that bring distributed R code closer to native Scala code)

Let's take a look at each requirement to ensure that you properly consider the trade-offs or benefits that they provide.

Installing R

Starting with the first requirement, the R runtime is expected to be preinstalled in *every* node in the cluster; this is a requirement specific to `spark_apply()`.

Failure to install R in every node will trigger a `Cannot run program, no such file or directory` error when you attempt to use `spark_apply()`.

Contact your cluster administrator to consider making the R runtime available throughout the entire cluster. If R is already installed, you can specify the installation path to use with the `spark.r.command` configuration setting, as shown here:

```
config <- spark_config()
config["spark.r.command"] <- "<path-to-r-version>"

sc <- spark_connect(master = "local", config = config)
sdf_len(sc, 10) %>% spark_apply(function(e) e)
```

A *homogeneous cluster* is required since the driver node distributes, and potentially compiles, packages to the workers. For instance, the driver and workers must have the same processor architecture, system libraries, and so on. This is usually the case for most clusters, but might not be true for yours.

Different cluster managers, Spark distributions, and cloud providers support different solutions to install additional software (like R) across every node in the cluster; follow instructions when installing R over each worker node. Here are a few examples:

Spark Standalone
Requires connecting to each machine and installing R; tools like `pssh` allow you to run a single installation command against multiple machines.

Cloudera
Provides an R parcel (see the Cloudera blog post "How to Distribute Your R code with sparklyr and Cloudera Data Science Workbench" (*http://bit.ly/33JAfu4*)), which enables R over each worker node.

Amazon EMR
R is preinstalled when starting an EMR cluster as mentioned in "Amazon" on page 104.

Microsoft HDInsight
R is preinstalled and no additional steps are needed.

Livy
Livy connections *do not* support distributing packages because the client machine where the libraries are precompiled might not have the same processor architecture or operating systems as the cluster machines.

Strictly speaking, this completes the last requirement for your cluster. However, we strongly recommend you use Apache Arrow with `spark_apply()` to support large-scale computation with minimal overhead.

Apache Arrow

Before introducing Apache Arrow, we'll discuss how data is stored and transferred between Spark and R. As R was designed from its inception to perform fast numeric computations to accomplish this, it's important to figure out the best way to store data.

Some computing systems store data internally by row; however, most interesting numerical operations require data to be processed by column. For example, calculating the mean of a column requires processing each column on its own, not the entire row. Spark stores data by default by row, since it's easier to partition; in contrast, R stores data by column. Therefore, something needs to transform both representations when data is transferred between Spark and R, as shown in Figure 11-6.

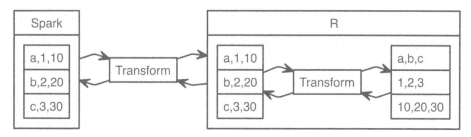

Figure 11-6. Data transformation between Spark and R

This transformation from rows to columns needs to happen for each partition. In addition, data also must be transformed from Scala's internal representation to R's internal representation. Apache Arrow reduces these transformations that waste a lot of CPU cycles.

Apache Arrow is a cross-language development platform for in-memory data. In Spark, it speeds up transferring data between Scala and R by defining a common data format compatible with many programming languages. Instead of having to transform between Scala's internal representation and R's, you can use the same structure for both languages. In addition, transforming data from row-based storage to columnar storage is performed in parallel in Spark, which can be further optimized by using the columnar storage formats presented in Chapter 8. The improved transformations are shown in Figure 11-7.

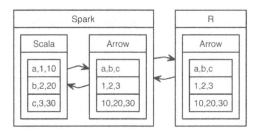

Figure 11-7. Data transformation between Spark and R using Apache Arrow

Apache Arrow is not required but is strongly recommended while you are working with spark_apply(). It has been available since Spark 2.3.0; however, it requires system administrators to install the Apache Arrow runtime in every node (see *http:// arrow.apache.org/install/*).

In addition, to use Apache Arrow with sparklyr, you also need to install the arrow package:

```
install.packages("arrow")
```

Before we use arrow, let's take a measurement to validate:

```
system.time(
  sdf_len(sc, 10^4) %>% spark_apply(nrow) %>% collect()
)
   user  system elapsed
  0.240   0.020   7.957
```

In our particular system, processing 10,000 rows takes about 8 seconds. To enable Arrow, simply include the library and use spark_apply() as usual. Let's measure how long it takes spark_apply() to process 1 million rows:

```
library(arrow)
system.time(
  sdf_len(sc, 10^6) %>% spark_apply(nrow) %>% collect()
)
   user  system elapsed
  0.317   0.021   3.922
```

In our system, Apache Arrow can process 100 times more data in half the time: just 4 seconds.

Most functionality in arrow simply works in the background, improving performance and data serialization; however, there is one setting you should be aware of. The spark.sql.execution.arrow.maxRecordsPerBatch configuration setting specifies the default size of each arrow data transfer. It's shared with other Spark components and defaults to 10,000 rows:

```
library(arrow)
sdf_len(sc, 2 * 10^4) %>% spark_apply(nrow)

# Source: spark<?> [?? x 1]
  result
   <int>
1  10000
2  10000
```

You might need to adjust this number based on how much data your system can handle, making it smaller for large datasets or bigger for operations that require records to be processed together. We can change this setting to 5,000 rows and verify the partitions change appropriately:

```
config <- spark_config()
config$spark.sql.execution.arrow.maxRecordsPerBatch <- 5 * 10^3

sc <- spark_connect(master = "local", config = config)
sdf_len(sc, 2 * 10^4) %>% spark_apply(nrow)

# Source: spark<?> [?? x 1]
  result
   <int>
1   5000
2   5000
3   5000
4   5000
```

So far we've presented use cases, main operations, and cluster requirements. Now we'll discuss the troubleshooting techniques useful when distributing R code.

Troubleshooting

A custom transformation can fail for many reasons. To learn how to troubleshoot errors, let's simulate a problem by triggering an error ourselves:

```
sdf_len(sc, 1) %>% spark_apply(~stop("force an error"))

Error in force(code) :
  sparklyr worker rscript failure, check worker logs for details
    Log: wm_bx4cn70s6h0r5vgsldm0000gn/T/Rtmpob83LD/file2aac1a6188_spark.log

---- Output Log ----
19/03/11 14:12:24 INFO sparklyr: Worker (1) completed wait using lock for RScript
```

Notice that the error message mentions inspecting the logs. When running in local mode, you can simply run the following:

```
spark_log(sc, filter = "terminated unexpectedly")

19/03/11 14:12:24 ERROR sparklyr: RScript (1) terminated unexpectedly:
                                  force an error
```

This points to the artificial `stop("force an error")` error we mentioned. However, if you're not working in local mode, you will have to retrieve the worker logs from your cluster manager. Since this can be cumbersome, one alternative is to rerun `spark_apply()` but return the error message yourself:

```
sdf_len(sc, 1) %>% spark_apply(~tryCatch(
    stop("force an error"),
    error = function(e) e$message
))
# Source: spark<?> [?? x 1]
  result
  <chr>
1 force an error
```

Among the other, more advanced troubleshooting techniques applicable to `spark_apply()`, the following sections present these techniques in order. You should try to troubleshoot by using worker logs first, then identifying partitioning errors, and finally, attempting to debug a worker node.

Worker Logs

Whenever `spark_apply()` is executed, information regarding execution is written over each worker node. You can use this log to write custom messages to help you diagnose and fine-tune your code.

For instance, suppose that we don't know what the first column name of `df` is. We can write a custom log message executed from the worker nodes using `worker_log()` as follows:

```
sdf_len(sc, 1) %>% spark_apply(function(df) {
  worker_log("the first column in the data frame is named ", names(df)[[1]])
  df
})
# Source: spark<?> [?? x 1]
      id
* <int>
1     1
```

When running locally, we can filter the log entries for the worker as follows:

```
spark_log(sc, filter = "sparklyr: RScript")

18/12/18 11:33:47 INFO sparklyr: RScript (3513) the first column in
the dataframe is named id
18/12/18 11:33:47 INFO sparklyr: RScript (3513) computed closure
18/12/18 11:33:47 INFO sparklyr: RScript (3513) updating 1 rows
18/12/18 11:33:47 INFO sparklyr: RScript (3513) updated 1 rows
18/12/18 11:33:47 INFO sparklyr: RScript (3513) finished apply
18/12/18 11:33:47 INFO sparklyr: RScript (3513) finished
```

Notice that the logs print our custom log entry, showing that id is the name of the first column in the given DataFrame.

This functionality is useful when troubleshooting errors; for instance, if we force an error using the stop() function:

```
sdf_len(sc, 1) %>% spark_apply(function(df) {
  stop("force an error")
})
```

We will get an error similar to the following:

```
Error in force(code) :
  sparklyr worker rscript failure, check worker logs for details
```

As suggested by the error, we can look in the worker logs for the specific errors as follows:

```
spark_log(sc)
```

This will show an entry containing the error and the call stack:

```
18/12/18 11:26:47 INFO sparklyr: RScript (1860) computing closure
18/12/18 11:26:47 ERROR sparklyr: RScript (1860) terminated unexpectedly:
                                  force an error
18/12/18 11:26:47 ERROR sparklyr: RScript (1860) collected callstack:
11: stop("force and error")
10: (function (df)
{
    stop("force and error")
})(structure(list(id = 1L), class = "data.frame", row.names = c(NA,
-1L)))
```

Notice that spark_log(sc) only retrieves the worker logs when you're using local clusters. When running in proper clusters with multiple machines, you will have to use the tools and user interface provided by the cluster manager to find these log entries.

Resolving Timeouts

When you are running with several hundred executors, it becomes more likely that some tasks will hang indefinitely. In this situation, most of the tasks in your job complete successfully, but a handful of them are still running and do not fail or succeed.

Suppose that you need to calculate the size of many web pages. You could use spark_apply() with something similar to:

```
sdf_len(sc, 3, repartition = 3) %>%
  spark_apply(~ download.file("https://google.com", "index.html") +
                file.size("index.html"))
```

Some web pages might not exist or take too long to download. In this case, most tasks will succeed, but a few will hang. To prevent these few tasks from blocking all computations, you can use the `spark.speculation` Spark setting. With this setting enabled, once 75% of all tasks succeed, Spark will look for tasks taking longer than the median task execution time and retry. You can use the `spark.speculation.multiplier` setting to configure the time multiplier used to determine when a task is running slow.

Therefore, for this example, you could configure Spark to retry tasks that take four times longer than the median as follows:

```
config <- spark_config()
config["spark.speculation"] <- TRUE
config["spark.speculation.multiplier"] <- 4
```

Inspecting Partitions

If a particular partition fails, you can detect the broken partition by computing a digest and then retrieving that particular partition. As usual, you can install `digest` from CRAN before connecting to Spark:

```
sdf_len(sc, 3) %>% spark_apply(function(x) {
    worker_log("processing ", digest::digest(x), " partition")
    # your code
    x
})
```

This will add an entry similar to:

```
18/11/03 14:48:32 INFO sparklyr: RScript (2566)
  processing f35b1c321df0162e3f914adfb70b5416 partition
```

When executing this in your cluster, look in the logs for the task that is not finishing. Once you have that digest, you can cancel the job. Then you can use that digest to retrieve the specific DataFrame to R with something like this:

```
sdf_len(sc, 3) %>% spark_apply(function(x) {
    if (identical(digest::digest(x),
                  "f35b1c321df0162e3f914adfb70b5416")) x else x[0,]
}) %>% collect()

# A tibble: 1 x 1
  result
   <int>
1      1
```

You can then run this in R to troubleshoot further.

Debugging Workers

A debugger is a tool that lets you execute your code line by line; you can use this to troubleshoot `spark_apply()` for local connections. You can start `spark_apply()` in debug mode by using the `debug` parameter and then following the instructions:

```
sdf_len(sc, 1) %>% spark_apply(function() {
  stop("Error!")
}, debug = TRUE)

Debugging spark_apply(), connect to worker debugging session as follows:
  1. Find the workers <sessionid> and <port> in the worker logs, from RStudio
     click 'Log' under the connection, look for the last entry with contents:
     'Session (<sessionid>) is waiting for sparklyr client to connect to
      port <port>'
  2. From a new R session run:
     debugonce(sparklyr:::spark_worker_main)
     sparklyr:::spark_worker_main(<sessionid>, <port>)
```

As these instructions indicate, you'll need to connect "as the worker node" from a different R session and then step through the code. This method is less straightforward than previous ones, since you'll also need to step through some lines of sparklyr code; thus, we only recommend this as a last resort. (You can also try the online resources described in Chapter 2.)

Let's now wrap up this chapter with a brief recap of the functionality we presented.

Recap

This chapter presented `spark_apply()` as an advanced technique you can use to fill gaps in functionality in Spark or its many extensions. We presented example use cases for `spark_apply()` to parse data, model in parallel many small datasets, perform a grid search, and call web APIs. You learned how partitions relate to `spark_apply()`, and how to create custom groups, distribute contextual information across all nodes, and troubleshoot problems, limitations, and cluster configuration caveats.

We also strongly recommended using Apache Arrow as a library when working with Spark with R, and presented installation, use cases, and considerations you should be aware of.

Up to this point, we've only worked with large datasets of static data, which doesn't change over time and remains invariant while we analyze, model, and visualize them. In Chapter 12, we will introduce techniques to process datasets that, in addition to being large, are also growing in such a way that they resemble a stream of information.

## CHAPTER 12

Streaming

Our stories aren't over yet.

—*Arya Stark*

Looking back at the previous chapters, we've covered a good deal, but not everything. We've analyzed tabular datasets, performed unsupervised learning over raw text, analyzed graphs and geographic datasets, and even transformed data with custom R code! So now what?

Though we weren't explicit about this, we've assumed until this point that your data is static, and didn't change over time. But suppose for a moment your job is to analyze traffic patterns to give recommendations to the department of transportation. A reasonable approach would be to analyze historical data and then design predictive models that compute forecasts overnight. Overnight? That's very useful, but traffic patterns change by the hour and even by the minute. You could try to preprocess and predict faster and faster, but eventually this model breaks—you can't load large-scale datasets, transform them, score them, unload them, and repeat this process by the second.

Instead, we need to introduce a different kind of dataset—one that is not static but rather dynamic, one that is like a table but is growing constantly. We will refer to such datasets as *streams*.

Overview

We know how to work with large-scale static datasets, but how can we reason about large-scale real-time datasets? Datasets with an infinite amount of entries are known as *streams*.

For static datasets, if we were to do real-time scoring using a pretrained topic model, the entries would be lines of text; for real-time datasets, we would perform the same scoring over an infinite number of lines of text. Now, in practice, you will never process an infinite number of records. You will eventually stop the stream—or this universe might end, whichever comes first. Regardless, thinking of the datasets as infinite makes it much easier to reason about them.

Streams are most relevant when processing real-time data—for example, when analyzing a Twitter feed or stock prices. Both examples have well-defined columns, like "tweet" or "price," but there are always new rows of data to be analyzed.

Spark Streaming provides scalable and fault-tolerant data processing over streams of data. That means you can use many machines to process multiple streaming sources, perform joins with other streams or static sources, and recover from failures with at-least-once guarantees (each message is certain to be delivered, but may do so multiple times).

In Spark, you create streams by defining a *source*, a *transformation*, and a *sink*; you can think of these steps as reading, transforming, and writing a stream, as shown in Figure 12-1.

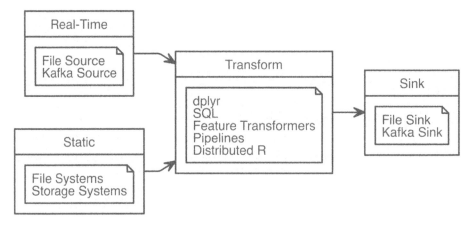

Figure 12-1. Working with Spark Streaming

Let's take a look at each of these a little more closely:

Reading
> Streams read data using any of the stream_read_*() functions; the read operation defines the *source* of the stream. You can define one or multiple sources from which to read.

Transforming
> A stream can perform one or multiple transformations using dplyr, SQL, feature transformers, scoring pipelines, or distributed R code. Transformations can not only be applied to one or more streams, but can also use a combination of

streams and static data sources; for instance, those loaded into Spark with spark_read_() functions—this means that you can combine static data and real-time data sources with ease.

Writing

The write operations are performed with the family of stream_write_*() functions, while the read operation defined the sink of the stream. You can specify a single sink or multiple ones to write data to.

You can read and write to streams in several different file formats: CSV, JSON, Parquet, Optimized Row Columnar (ORC), and text (see Table 12-1). You also can read and write from and to Kafka, which we will introduce later on.

Table 12-1. Spark functions to read and write streams

Format	Read	Write
CSV	stream_read_csv	stream_write_csv
JSON	stream_read_json	stream_write_json
Kafka	stream_read_kafka	stream_write_kafka
ORC	stream_read_orc	stream_write_orc
Parquet	stream_read_parquet	stream_write_parquet
Text	stream_read_text	stream_write_text
Memory		stream_write_memory

Since the transformation step is optional, the simplest stream we can define is one that continuously copies text files between source and destination.

First, install the future package using install.packages("future") and connect to Spark.

Since a stream requires the source to exist, create a _source_ folder:

```
dir.create("source")
```

We are now ready to define our first stream!

```
stream <- stream_read_text(sc, "source/") %>%
  stream_write_text("destination/")
```

The streams starts running with stream_write_*(); once executed, the stream will monitor the `source` path and process data into the *destination* / path as it arrives.

We can use stream_generate_test() to produce a file every second containing lines of text that follow a given distribution; you can read more about this in Appendix A. In practice, you would connect to existing sources without having to generate data artificially. We can then use view_stream() to track the rows per second (rps) being processed in the source, and in the destination, and their latest values over time:

```
future::future(stream_generate_test(interval = 0.5))

stream_view(stream)
```

The result is shown in Figure 12-2.

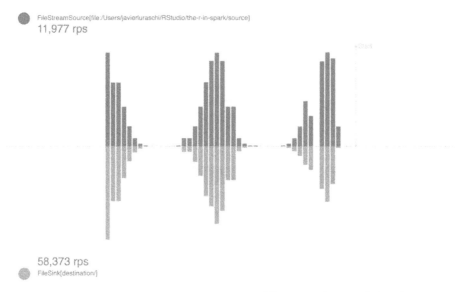

Figure 12-2. Monitoring a stream generating rows following a binomial distribution

Notice that the rps rate in the destination stream is higher than that in the source stream. This is expected and desirable since Spark measures incoming rates from the source stream, but also actual row-processing times in the destination stream. For example, if 10 rows per second are written to the *source/* path, the incoming rate is 10 rps. However, if it takes Spark only 0.01 seconds to write all those 10 rows, the output rate is 100 rps.

Use stream_stop() to properly stop processing data from this stream:

```
stream_stop(stream)
```

This exercise introduced how we can easily start a Spark stream that reads and writes data based on a simulated stream. Let's do something more interesting than just copying data with proper transformations.

Transformations

In a real-life scenario, the incoming data from a stream would not be written as is to the output. The Spark Streaming job would make transformations to the data, and then write the transformed data.

Streams can be transformed using `dplyr`, SQL queries, ML pipelines, or R code. We can use as many transformations as needed in the same way that Spark DataFrames can be transformed with `sparklyr`.

The source of the transformation can be a stream or DataFrame, but the output is always a stream. If needed, you can always take a snapshot from the destination stream and then save the output as a DataFrame. That is what `sparklyr` will do for you if a destination stream is not specified.

Each of the following subsections covers an option provided by `sparklyr` to perform transformations on a stream.

Analysis

You can analyze streams with `dplyr` verbs and SQL using `DBI`. As a quick example, we will filter rows and add columns over a stream. We won't explicitly call `stream_gener ate_test()`, but you can call it on your own through the `later` package if you feel the urge to verify that data is being processed continuously:

```
library(dplyr)

stream_read_csv(sc, "source") %>%
  filter(x > 700) %>%
  mutate(y = round(x / 100))

# Source: spark<?> [inf x 2]
        x      y
    <int>  <dbl>
 1    701      7
 2    702      7
 3    703      7
 4    704      7
 5    705      7
 6    706      7
 7    707      7
 8    708      7
 9    709      7
10    710      7
# … with more rows
```

It's also possible to perform aggregations over the entire history of the stream. The history could be filtered or not:

```
stream_read_csv(sc, "source") %>%
  filter(x > 700) %>%
  mutate(y = round(x / 100)) %>%
  count(y)

# Source: spark<?> [inf x 2]
      y      n
  <dbl>  <dbl>
```

```
1       8 25902
2       9 25902
3      10 13210
4       7 12692
```

Grouped aggregations of the latest data in the stream require a timestamp. The time-stamp will note when the reading function (in this case stream_read_csv()) first "saw" that specific record. In Spark Streaming terminology, the timestamp is called a *watermark*. The spark_watermark() function adds the timestamp. In this example, the watermark will be the same for all records, since the five files were read by the stream after they were created. Note that only Kafka and memory *outputs* support watermarks:

```
stream_read_csv(sc, "source") %>%
  stream_watermark()

# Source: spark<?> [inf x 2]
        x timestamp
    <int> <dttm>
 1    276 2019-06-30 07:14:21
 2    277 2019-06-30 07:14:21
 3    278 2019-06-30 07:14:21
 4    279 2019-06-30 07:14:21
 5    280 2019-06-30 07:14:21
 6    281 2019-06-30 07:14:21
 7    282 2019-06-30 07:14:21
 8    283 2019-06-30 07:14:21
 9    284 2019-06-30 07:14:21
10    285 2019-06-30 07:14:21
# … with more rows
```

After the watermark is created, you can use it in the group_by() verb. You can then pipe it into a summarise() function to get some stats of the stream:

```
stream_read_csv(sc, "source") %>%
  stream_watermark() %>%
  group_by(timestamp) %>%
  summarise(
    max_x = max(x, na.rm = TRUE),
    min_x = min(x, na.rm = TRUE),
    count = n()
  )

# Source: spark<?> [inf x 4]
  timestamp             max_x min_x  count
  <dttm>                <int> <int>  <dbl>
1 2019-06-30 07:14:55    1000     1 259332
```

Modeling

Spark streams currently don't support training on real-time datasets. Aside from the technical challenges, even if it were possible, it would be quite difficult to train models since the model itself would need to adapt over time. Known as *online learning*, this is perhaps something that Spark will support in the future.

That said, there are other modeling concepts we can use with streams, like feature transformers and scoring. Let's try out a feature transformer with streams, and leave scoring for the next section, since we will need to train a model.

The next example uses the `ft_bucketizer()` feature transformer to modify the stream followed by regular `dplyr` functions, which you can use just as you would with static datasets:

```
stream_read_csv(sc, "source") %>%
  mutate(x = as.numeric(x)) %>%
  ft_bucketizer("x", "buckets", splits = 0:10 * 100) %>%
  count(buckets)  %>%
  arrange(buckets)
```

```
# Source:     spark<?> [inf x 2]
# Ordered by: buckets
    buckets      n
      <dbl> <dbl>
1         0 25747
2         1 26008
3         2 25992
4         3 25908
5         4 25905
6         5 25903
7         6 25904
8         7 25901
9         8 25902
10        9 26162
```

Pipelines

Spark pipelines can be used for scoring streams, but not to train over streaming data. The former is fully supported, while the latter is a feature under active development by the Spark community.

To score a stream, it's necessary to first create our model. So let's build, fit, and save a simple pipeline:

```
cars <- copy_to(sc, mtcars)

model <- ml_pipeline(sc) %>%
  ft_binarizer("mpg", "over_30", 30) %>%
  ft_r_formula(over_30 ~ wt) %>%
  ml_logistic_regression() %>%
  ml_fit(cars)
```

If you choose to, you can make use of other concepts presented in Chapter 5, like saving and reloading pipelines through ml_save() and ml_load() before scoring streams.

We can then generate a stream based on mtcars using stream_generate_test(), and score the model using ml_transform():

```
future::future(stream_generate_test(mtcars, "cars-stream", iterations = 5))

ml_transform(model, stream_read_csv(sc, "cars-stream"))
# Source: spark<?> [inf x 17]
      mpg   cyl  disp    hp  drat    wt  qsec    vs    am  gear  carb over_30
    <dbl> <int> <dbl> <int> <dbl> <dbl> <dbl> <int> <int> <int> <int>   <dbl>
 1   15.5     8  318    150  2.76  3.52  16.9     0     0     3     2       0
 2   15.2     8  304    150  3.15  3.44  17.3     0     0     3     2       0
 3   13.3     8  350    245  3.73  3.84  15.4     0     0     3     4       0
 4   19.2     8  400    175  3.08  3.84  17.0     0     0     3     2       0
 5   27.3     4   79     66  4.08  1.94  18.9     1     1     4     1       0
 6   26       4  120.    91  4.43  2.14  16.7     0     1     5     2       0
 7   30.4     4   95.1  113  3.77  1.51  16.9     1     1     5     2       1
 8   15.8     8  351    264  4.22  3.17  14.5     0     1     5     4       0
 9   19.7     6  145    175  3.62  2.77  15.5     0     1     5     6       0
10   15       8  301    335  3.54  3.57  14.6     0     1     5     8       0
# … with more rows, and 5 more variables: features <list>, label <dbl>,
#    rawPrediction <list>, probability <list>, prediction <dbl>
```

Though this example was put together with a few lines of code, what we just accomplished is actually quite impressive. You copied data into Spark, performed feature engineering, trained a model, and scored the model over a real-time dataset, with just seven lines of code! Let's try now to use custom transformations, in real time.

Distributed R

Arbitrary R code can also be used to transform a stream with the use of spark_apply(). This approach follows the same principles discussed in Chapter 11, where spark_apply() runs R code over each executor in the cluster where data is available. This enables processing high-throughput streams and fulfills low-latency requirements:

```
stream_read_csv(sc, "cars-stream") %>%
  select(mpg) %>%
  spark_apply(~ round(.x), mpg = "integer") %>%
  stream_write_csv("cars-round")
```

which, as you would expect, processes data from cars-stream into cars-round by running the custom round() R function. Let's peek into the output sink:

```
spark_read_csv(sc, "cars-round")
```

```
# Source: spark<carsround> [?? x 1]
      mpg
    <dbl>
 1    16
 2    15
 3    13
 4    19
 5    27
 6    26
 7    30
 8    16
 9    20
10    15
# … with more rows
```

Again, make sure you apply the concepts you already know about `spark_apply()` when using streams; for instance, you should consider using `arrow` to significantly improve performance. Before we move on, disconnect from Spark:

```
spark_disconnect(sc)
```

This was our last transformation for streams. We'll now learn how to use Spark Streaming with Kafka.

Kafka

Apache Kafka is an open source stream-processing software platform developed by LinkedIn and donated to the Apache Software Foundation. It is written in Scala and Java. To describe it using an analogy, Kafka is to real-time storage what Hadoop is to static storage.

Kafka stores the stream as records, which consist of a key, a value, and a timestamp. It can handle multiple streams that contain different information, by categorizing them by topic. Kafka is commonly used to connect multiple real-time applications. A *producer* is an application that streams data into Kafka, while a *consumer* is the one that reads from Kafka; in Kafka terminology, a consumer application *subscribes* to topics. Therefore, the most basic workflow we can accomplish with Kafka is one with a single producer and a single consumer; this is illustrated in Figure 12-3.

Figure 12-3. A basic Kafka workflow

If you are new to Kafka, we don't recommend you run the code from this section. However, if you're really motivated to follow along, you will first need to install Kafka as explained in Appendix A or deploy it in your cluster.

Using Kafka also requires you to have the Kafka package when connecting to Spark. Make sure this is specified in your connection `config`:

```
library(sparklyr)
library(dplyr)

sc <- spark_connect(master = "local",version = "2.3", config = list(
  sparklyr.shell.packages = "org.apache.spark:spark-sql-kafka-0-10_2.11:2.3.1"
))
```

Once connected, it's straightforward to read data from a stream:

```
stream_read_kafka(
  sc,
  options = list(
    kafka.bootstrap.server = "host1:9092, host2:9092",
    subscribe = "<topic-name>"
    )
  )
```

However, notice that you need to properly configure the `options` list; `kafka.boot strap.server` expects a list of Kafka hosts, while `topic` and `subscribe` define which topic should be used when writing or reading from Kafka, respectively.

Though we've started by presenting a simple single-producer and single-consumer use case, Kafka also allows much more complex interactions. We will next read from one topic, process its data, and then write the results to a different topic. Systems that are producers and consumers from the same topic are referred to as *stream processors*. In Figure 12-4, the stream processor reads topic A and then writes results to topic B. This allows for a given consumer application to read results instead of "raw" feed data.

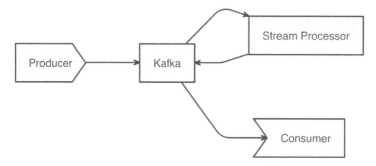

Figure 12-4. A Kafka workflow using stream processors

Three modes are available when processing Kafka streams in Spark: *complete*, *update*, and *append*. The `complete` mode provides the totals for every group every time there is a new batch; `update` provides totals for only the groups that have updates in the latest batch; and `append` adds raw records to the target topic. The `append` mode is not meant for aggregates, but works well for passing a filtered subset to the target topic.

In our next example, the producer streams random letters into Kafka under a `letters` topic. Then, Spark will act as the stream processor, reading the `letters` topic and computing unique letters, which are then written back to Kafka under the `totals` topic. We'll use the `update` mode when writing back into Kafka; that is, only the totals that changed will be sent to Kafka. This change is determined after each batch from the `letters` topic:

```
hosts  <- "localhost:9092"

read_options <- list(kafka.bootstrap.servers = hosts, subscribe = "letters")
write_options <- list(kafka.bootstrap.servers = hosts, topic = "totals")

stream_read_kafka(sc, options = read_options) %>%
  mutate(value = as.character(value)) %>%         # coerce into a character
  count(value) %>%                                # group and count letters
  mutate(value = paste0(value, "=", n)) %>%       # kafka expects a value field
  stream_write_kafka(mode = "update",
                     options = write_options)
```

You can take a quick look at totals by reading from Kafka:

```
stream_read_kafka(sc, options = totals_options)
```

Using a new terminal session, use Kafka's command-line tool to manually add single letters into the `letters` topic:

```
kafka-console-producer.sh --broker-list localhost:9092 --topic letters
>A
>B
>C
```

The letters that you input are pushed to Kafka, read by Spark, aggregated within Spark, and pushed back into Kafka, Then, finally, they are consumed by Spark again to give you a glimpse into the `totals` topic. This was quite a setup, but also a realistic configuration commonly found in real-time processing projects.

Next, we will use the Shiny framework to visualize streams, in real time!

Shiny

Shiny's reactive framework is well suited to support streaming information, which you can use to display real-time data from Spark using `reactiveSpark()`. There is far more to learn about Shiny than we could possibly present here. However, if you're already familiar with Shiny, this example should be quite easy to understand.

We have a modified version of the *k*-means Shiny example that, instead of getting the data from the static `iris` dataset, is generated with `stream_generate_test()`, consumed by Spark, retrieved to Shiny through `reactiveSpark()`, and then displayed as shown in Figure 12-5.

To run this example, store the following Shiny app under `shiny/shiny-stream.R`:

```
library(sparklyr)
library(shiny)

unlink("shiny-stream", recursive = TRUE)
dir.create("shiny-stream", showWarnings = FALSE)

sc <- spark_connect(
  master = "local", version = "2.3",
  config = list(sparklyr.sanitize.column.names = FALSE))

ui <- pageWithSidebar(
  headerPanel('Iris k-means clustering from Spark stream'),
  sidebarPanel(
    selectInput('xcol', 'X Variable', names(iris)),
    selectInput('ycol', 'Y Variable', names(iris),
                selected=names(iris)[[2]]),
    numericInput('clusters', 'Cluster count', 3,
                 min = 1, max = 9)
  ),
  mainPanel(plotOutput('plot1'))
)

server <- function(input, output, session) {
  iris <- stream_read_csv(sc, "shiny-stream",
                          columns = sapply(datasets::iris, class)) %>%
    reactiveSpark()

  selectedData <- reactive(iris()[, c(input$xcol, input$ycol)])
  clusters <- reactive(kmeans(selectedData(), input$clusters))

  output$plot1 <- renderPlot({
    par(mar = c(5.1, 4.1, 0, 1))
    plot(selectedData(), col = clusters()$cluster, pch = 20, cex = 3)
    points(clusters()$centers, pch = 4, cex = 4, lwd = 4)
  })
}

shinyApp(ui, server)
```

This Shiny application can then be launched with `runApp()`, like so:

```
shiny::runApp("shiny/shiny-stream.R")
```

While the Shiny app is running, launch a new R session from the same directory and create a test stream with `stream_generate_test()`. This will generate a stream of continuous data that Spark can process and Shiny can visualize (as illustrated in Figure 12-5):

```
sparklyr::stream_generate_test(datasets::iris, "shiny/shiny-stream",
                               rep(5, 10^3))
```

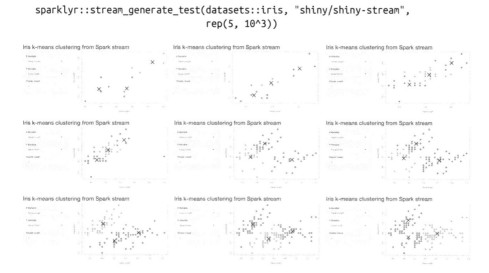

Figure 12-5. Progression of Spark reactive loading data into the Shiny app

In this section you learned how easy it is to create a Shiny app that can be used for several different purposes, such as monitoring and dashboarding.

In a more complex implementation, the source would more likely be a Kafka stream.

Before we transition, disconnect from Spark and clear the folders that we used:

```
spark_disconnect(sc)

unlink(c("source", "destination", "cars-stream",
         "car-round", "shiny/shiny-stream"), recursive = TRUE)
```

Recap

From static datasets to real-time datasets, you've truly mastered many of the large-scale computing techniques. Specifically, in this chapter, you learned how static data can be generalized to real time if we think of it as an infinite table. We were then able to create a simple stream, without any data transformations, that copies data from point A to point B.

This humble start became quite useful when you learned about the several different transformations you can apply to streaming data—from data analysis transformations using the `dplyr` and `DBI` packages, to feature transformers introduced while modeling, to fully fledged pipelines capable of scoring in real time, to, last but not least, transforming datasets with custom R code. This was a lot to digest, for sure.

We then presented Apache Kafka as a reliable and scalable solution for real-time data. We showed you how a real-time system could be structured by introducing you to consumers, producers, and topics. These, when properly combined, create powerful abstractions to process real-time data.

Then we closed with "a cherry on top of the sundae": presenting how to use Spark Streaming in Shiny. Since a stream can be transformed into a reactive (which is the lingua franca of the world of reactivity), the ease of this approach was a nice surprise.

It's time now to move on to our very last (and quite short) chapter, Chapter 13; there we'll try to persuade you to use your newly acquired knowledge for the benefit of the Spark and R communities at large.

Contributing

Hold the door, hold the door.

—Hodor

In Chapter 12, we equipped you with the tools to tackle large-scale and real-time data processing in Spark using R. In this final chapter we focus less on learning and more on giving back to the Spark and R communities or colleagues in your professional career. It really takes an entire community to keep this going, so we are counting on you!

There are many ways to contribute, from helping community members and opening GitHub issues to providing new functionality for yourself, colleagues, or the R and Spark community. However, we'll focus here on writing and sharing code that extends Spark, to help others use new functionality you can provide as an author of Spark extensions using R. Specifically, you'll learn what an extension is, the different types of extensions you can build, what building tools are available, and when and how to build an extension from scratch.

You will also learn how to make use of the hundreds of extensions available in Spark and the millions of components available in Java that can easily be used in R. We'll also cover how to create code natively in Scala that makes use of Spark. As you might know, R is a great language for interfacing with other languages, such as C++, SQL, Python, and many others. It's no surprise, then, that working with Scala from R will follow similar practices that make R ideal for providing easy-to-use interfaces that make data processing productive and that are loved by many of us.

Overview

When you think about giving back to your larger coding community, the most important question you can ask about any piece of code you write is: would this code be useful to someone else?

Let's start by considering one of the first and simplest lines of code presented in this book. This code was used to load a simple CSV file:

```
spark_read_csv(sc, "cars.csv")
```

Code this basic is probably not useful to someone else. However, you could tailor that same example to something that generates more interest, perhaps the following:

```
spark_read_csv(sc, "/path/that/is/hard/to/remember/data.csv")
```

This code is quite similar to the first. But if you work with others who are working with this dataset, the answer to the question about usefulness would be yes—this would very likely be useful to someone else!

This is surprising since it means that not all useful code needs to be advanced or complicated. However, for it to be useful to others, it does need to be packaged, presented, and shared in a format that is easy to consume.

A first attempt would be to save this into a *teamdata.R* file and write a function wrapping it:

```
load_team_data <- function() {
  spark_read_text(sc, "/path/that/is/hard/to/remember/data.csv")
}
```

This is an improvement, but it would require users to manually share this file again and again. Fortunately, this problem is easily solved in R, through *R packages*.

An R package contains R code in a format that is installable using the function `install.packages()`. One example is `sparklyr`. There are many other R packages available; you can also create your own. For those of you new to creating them, we encourage you to read Hadley Wickham's book, *R Packages* (O'Reilly). Creating an R package allows you to easily share your functions with others by sharing the package file in your organization.

Once you create a package, there are many ways of sharing it with colleagues or the world. For instance, for packages meant to be private, consider using Drat (*http://bit.ly/2N8J1f7*) or products like RStudio Package Manager (*http://bit.ly/2H5k807*). R packages meant for public consumption are made available to the R community in CRAN (*https://cran.r-project.org/*) (the Comprehensive R Archive Network).

These repositories of R packages allow users to install packages through `install.packages("teamdata")` without having to worry where to download the package from. It also allows other packages to reuse your package.

In addition to using R packages like `sparklyr`, `dplyr`, `broom`, and others to create new R packages that extend Spark, you can also use all the functionality available in the Spark API or Spark Extensions or write custom Scala code.

For instance, suppose that there is a new file format similar to a CSV but not quite the same. We might want to write a function named `spark_read_file()` that would take a path to this new file type and read it in Spark. One approach would be to use `dplyr` to process each line of text or any other R library using `spark_apply()`. Another would be to use the Spark API to access methods provided by Spark. A third approach would be to check whether someone in the Spark community has already provided a Spark extension that supports this new file format. Last but not least, you could write your own custom Scala code that makes use of any Java library, including Spark and its extensions. This is shown in Figure 13-1.

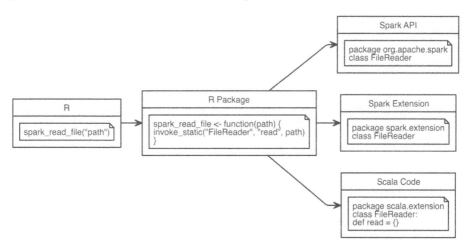

Figure 13-1. Extending Spark using the Spark API or Spark extensions, or writing Scala code

We will focus first on extending Spark using the Spark API, since the techniques required to call the Spark API are also applicable while calling Spark extensions or custom Scala code.

The Spark API

Before we introduce the Spark API, let's consider a simple and well-known problem. Suppose we want to count the number of lines in a distributed and potentially large text file—say, *cars.csv*, which we initialize as follows:

```
library(sparklyr)
library(dplyr)
sc <- spark_connect(master = "local")

cars <- copy_to(sc, mtcars)
spark_write_csv(cars, "cars.csv")
```

Now, to count how many lines are available in this file, we can run the following:

```
spark_read_text(sc, "cars.csv") %>% count()

# Source: spark<?> [?? x 1]
      n
  <dbl>
1    33
```

Easy enough: we used spark_read_text() to read the entire text file, followed by counting lines using dplyr's count(). Now, suppose that neither spark_read_text(), nor dplyr, nor any other Spark functionality, is available to you. How would you ask Spark to count the number of rows in *cars.csv*?

If you do this in Scala, you find in the Spark documentation that by using the Spark API you can count lines in a file as follows:

```
val textFile = spark.read.textFile("cars.csv")
textFile.count()
```

So, to use the functionality available in the Spark API from R, like spark.read.text File, you can use invoke(), invoke_static(), or invoke_new(). (As their names suggest, the first invokes a method from an object, the second invokes a method from a static object, and the third creates a new object.) We then use these functions to call Spark's API and execute code similar to the one provided in Scala:

```
spark_context(sc) %>%
  invoke("textFile", "cars.csv", 1L) %>%
  invoke("count")

[1] 33
```

While the invoke() function was originally designed to call Spark code, it can also call any code available in Java. For instance, we can create a Java BigInteger with the following code:

```
invoke_new(sc, "java.math.BigInteger", "1000000000")
```

```
<jobj[225]>
  java.math.BigInteger
  1000000000
```

As you can see, the object created is not an R object but rather a proper Java object. In R, this Java object is represented by the `spark_jobj`. These objects are meant to be used with the `invoke()` functions or `spark_dataframe()` and `spark_connection()`. `spark_dataframe()` transforms a `spark_jobj` into a Spark DataFrame when possible, whereas `spark_connect()` retrieves the original Spark connection object, which can be useful to avoid passing the `sc` object across functions.

While calling the Spark API can be useful in some cases, most of the functionality available in Spark is already supported in `sparklyr`. Therefore, a more interesting way to extend Spark is by using one of its many existing extensions.

Spark Extensions

Before we get started with this section, consider navigating to spark-packages.org (*https://spark-packages.org/*), a site that tracks Spark extensions provided by the Spark community. Using the same techniques presented in the previous section, you can use these extensions from R.

For instance, there is Apache Solr (*http://bit.ly/2MmBfim*), a system designed to perform full-text searches over large datasets, something that Apache Spark currently does not support natively. Also, as of this writing, there is no extension for R to support Solr. So, let's try to solve this by using a Spark extension.

First, if you search "spark-packages.org" to find a Solr extension, you should be able to locate `spark-solr` (*http://bit.ly/2YQnwXw*). The "How to" extension mentions that the `com.lucidworks.spark:spark-solr:2.0.1` should be loaded. We can accomplish this in R using the `sparklyr.shell.packages` configuration option:

```
config <- spark_config()
config["sparklyr.shell.packages"] <- "com.lucidworks.spark:spark-solr:3.6.3"
config["sparklyr.shell.repositories"] <-
  "http://repo.spring.io/plugins-release/,http://central.maven.org/maven2/"

sc <- spark_connect(master = "local", config = config)
```

While specifying the `sparklyr.shell.packages` parameter is usually enough, for this particular extension, dependencies failed to download from the Spark packages repository. You would need to manually find the failed dependencies in the Maven repo (*https://mvnrepository.com*) and add further repositories under the `spar klyr.shell.repositories` parameter.

When you use an extension, Spark connects to the Maven package repository to retrieve it. This can take significant time depending on the extension and your download speeds. In this case, consider using the sparklyr.connect.timeout configuration parameter to allow Spark to download the required files.

From the spark-solr documentation, you would find that this extension can be used with the following Scala code:

```
val options = Map(
  "collection" -> "{solr_collection_name}",
  "zkhost" -> "{zk_connect_string}"
)

val df = spark.read.format("solr")
  .options(options)
  .load()
```

We can translate this to R code:

```
spark_session(sc) %>%
  invoke("read") %>%
  invoke("format", "solr") %>%
  invoke("option", "collection", "<collection>") %>%
  invoke("option", "zkhost", "<host>") %>%
  invoke("load")
```

This code will fail, however, since it would require a valid Solr instance and configuring Solr goes beyond the scope of this book. But this example provides insights as to how you can create Spark extensions. It's also worth mentioning that you can use spark_read_source() to read from generic sources to avoid writing custom invoke() code.

As pointed out in "Overview" on page 240, consider sharing code with others using R packages. While you could require users of your package to specify spar klyr.shell.packages, you can avoid this by registering dependencies in your R package. Dependencies are declared under a spark_dependencies() function; thus, for the example in this section:

```
spark_dependencies <- function(spark_version, scala_version, ...) {
  spark_dependency(
    packages = "com.lucidworks.spark:spark-solr:3.6.3",
    repositories = c(
      "http://repo.spring.io/plugins-release/",
      "http://central.maven.org/maven2/")
  )
}

.onLoad <- function(libname, pkgname) {
```

```
      sparklyr::register_extension(pkgname)
  }
```

The onLoad function is automatically called by R when your library loads. It should call register_extension(), which will then call back spark_dependencies(), to allow your extension to provide additional dependencies. This example supports Spark 2.4, but you should also support mapping Spark and Scala versions to the correct Spark extension version.

There are about 450 Spark extensions you can use; in addition, you can also use any Java library from a Maven repository (*http://bit.ly/2Mp0wrR*), where Maven Central has over 3 million artifacts. While not all Maven Central libraries might be relevant to Spark, the combination of Spark extensions and Maven repositories certainly opens many interesting possibilities for you to consider!

However, for those cases where no Spark extension is available, the next section will teach you how to use custom Scala code from your own R package.

Using Scala Code

Scala code enables you to use any method in the Spark API, Spark extensions, or Java library. In addition, writing Scala code when running in Spark can provide performance improvements over R code using spark_apply(). In general, the structure of your R package will contain R code and Scala code; however, the Scala code will need to be compiled as JARs (Java ARchive files) and included in your package. Figure 13-2 shows this structure.

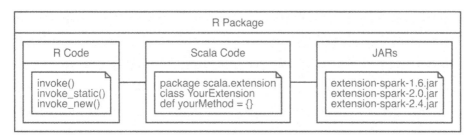

Figure 13-2. R package structure when using Scala code

As usual, the R code should be placed under a top-level *R* folder, Scala code under a *java* folder, and the compiled JARs under an *inst/java* folder. Though you are certainly welcome to manually compile the Scala code, you can use helper functions to download the required compiler and compile Scala code.

To compile Scala code, you'll need to install the Java Development Kit 8 (JDK8, for short). Download the JDK from Oracle's Java download page (*http://bit.ly/2P2UkYM*); this will require you to restart your R session.

You'll also need a Scala compiler for Scala 2.11 and 2.12 (*https://www.scala-lang.org/*). The Scala compilers can be automatically downloaded and installed using `download_scalac()`:

```
download_scalac()
```

Next, you'll need to compile your Scala sources using `compile_package_jars()`. By default, it uses `spark_compilation_spec()`, which compiles your sources for the following Spark versions:

```
## [1] "1.5.2" "1.6.0" "2.0.0" "2.3.0" "2.4.0"
```

You can also customize this specification by creating custom entries with `spark_compilation_spec()`.

While you could create the project structure for the Scala code from scratch, you can also simply call *spark_extension(path)* to create an extension in the given path. This extension will be mostly empty but will contain the appropriate project structure to call the Scala code.

Since `spark_extension()` is registered as a custom project extension in RStudio, you can also create an R package that extends Spark using Scala code from the File menu; click New Project and then select "R Package using Spark" as shown in Figure 13-3.

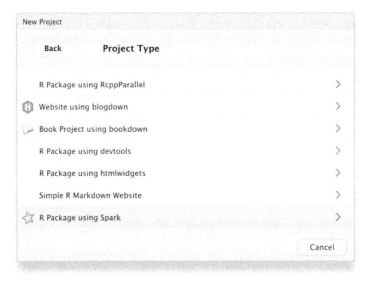

Figure 13-3. Creating a Scala extension package from RStudio

Once you are ready to compile your package JARs, you can simply run the following:

```
compile_package_jars()
```

Since the JARs are compiled by default into the *inst/* package path, when you are building the R package all the JARs will also be included within the package. This means that you can share or publish your R package, and it will be fully functional by R users. For advanced Spark users with most of their expertise in Scala, it's compelling to consider writing libraries for R users and the R community in Scala and then easily packaging these into R packages that are easy to consume, use, and share among them.

If you are interested in developing a Spark extension with R and you get stuck along the way, consider joining the `sparklyr` Gitter channel (*http://bit.ly/33ESccY*), where many of us hang out to help this wonderful community to grow. We hope to hear from you soon!

Recap

This final chapter introduced you to a new set of tools to use to expand Spark functionality beyond what R and R packages currently support. This vast new space of libraries includes more than 450 Spark extensions and millions of Java artifacts you can use in Spark from R. Beyond these resources, you also learned how to build Java artifacts using Scala code that can be easily embedded and compiled from R.

This brings us back to this book's purpose, presented early on; while we know you've learned how to perform large-scale computing using Spark in R, we're also confident that you have acquired the knowledge required to help other community members through Spark extensions. We can't wait to see your new creations, which will surely help grow the Spark and R communities at large.

To close, we hope the first chapters gave you an easy introduction to Spark and R. Following that, we presented analysis and modeling as foundations for using Spark, with the familiarity of R packages you already know and love. You moved on to learning how to perform large-scale computing in proper Spark clusters. The last third of this book focused on advanced topics: using extensions, distributing R code, processing real-time data, and, finally, giving back by using Spark extensions using R and Scala code.

We've tried to present the best possible content. However, if you know of any way to improve this book, please open a GitHub issue under the the-r-in-spark (*http://bit.ly/2HdkIZQ*) repo, and we'll address your suggestions in upcoming revisions. We hope you enjoyed reading this book, and that you've learned as much as we have learned while writing it. We hope it was worthy of your time—it has been an honor having you as our reader.

Supplemental Code References

Throughout the book we've included references to this appendix. Here we've included important content (and listed the sections where this material can be found).

Preface

Formatting

The following `ggplot2` theme was used to format all plots in this book:

```
plot_style <- function() {
  font <- "Helvetica"

  ggplot2::theme_classic() +
  ggplot2::theme(
    plot.title = ggplot2::element_text(
      family = font, size=14, color = "#222222"),
    plot.subtitle = ggplot2::element_text(
      family=font, size=12, color = "#666666"),

    legend.position = "right",
    legend.background = ggplot2::element_blank(),
    legend.title = ggplot2::element_blank(),
    legend.key = ggplot2::element_blank(),
    legend.text = ggplot2::element_text(
      family=font, size=14, color="#222222"),

    axis.title.y = ggplot2::element_text(
      margin = ggplot2::margin(t = 0, r = 8, b = 0, l = 0),
      size = 14, color="#666666"),
    axis.title.x = ggplot2::element_text(
      margin = ggplot2::margin(t = -2, r = 0, b = 0, l = 0),
      size = 14, color = "#666666"),
```

```
    axis.text = ggplot2::element_text(
      family=font, size=14, color="#222222"),
    axis.text.x = ggplot2::element_text(
      margin = ggplot2::margin(5, b = 10)),
    axis.ticks = ggplot2::element_blank(),
    axis.line = ggplot2::element_blank(),

    panel.grid.minor = ggplot2::element_blank(),
    panel.grid.major.y = ggplot2::element_line(color = "#eeeeee"),
    panel.grid.major.x = ggplot2::element_line(color = "#ebebeb"),

    panel.background = ggplot2::element_blank(),

    strip.background = ggplot2::element_rect(fill = "white"),
    strip.text = ggplot2::element_text(size  = 20,  hjust = 0)
  )
}
```

You can activate this with the following:

```
ggplot2::theme_set(plot_style())
```

Chapter 1

The World's Capacity to Store Information

The following script was used to generate Figure 1-1:

```
library(ggplot2)
library(dplyr)
library(tidyr)
read.csv("data/01-worlds-capacity-to-store-information.csv", skip = 8) %>%
  gather(key = storage, value = capacity, analog, digital) %>%
  mutate(year = X, terabytes = capacity / 1e+12) %>%
  ggplot(aes(x = year, y = terabytes, group = storage)) +
    geom_line(aes(linetype = storage)) +
    geom_point(aes(shape = storage)) +
    scale_y_log10(
      breaks = scales::trans_breaks("log10", function(x) 10^x),
      labels = scales::trans_format("log10", scales::math_format(10^x))
    ) +
    theme_light() +
    theme(legend.position = "bottom")
```

Daily Downloads of CRAN Packages

Figure 1-6 was generated through the following code:

```
downloads_csv <- "data/01-intro-r-cran-downloads.csv"
if (!file.exists(downloads_csv)) {
  downloads <- cranlogs::cran_downloads(from = "2014-01-01", to = "2019-01-01")
  readr::write_csv(downloads, downloads_csv)
```

```
}

cran_downloads <- readr::read_csv(downloads_csv)

ggplot(cran_downloads, aes(date, count)) +
  labs(title = "CRAN Packages",
       subtitle = "Total daily downloads over time") +
  geom_point(colour="black", pch = 21, size = 1) +
  scale_x_date() + xlab("year") + ylab("downloads") +
  scale_x_date(date_breaks = "1 year",
               labels = function(x) substring(x, 1, 4)) +
  scale_y_continuous(
     limits = c(0, 3.5 * 10^6),
     breaks = c(0.5 * 10^6, 10^6, 1.5 * 10^6, 2 * 10^6, 2.5 * 10^6, 3 * 10^6,
     3.5 * 10^6),
     labels = c("", "1M", "", "2M", "", "3M", "")
   )
```

Chapter 2

Prerequisites

Installing R

Download the R installer (*https://r-project.org/*) (see Figure A-1) and launch it for your Windows, Mac, or Linux platform.

The R Project for Statistical Computing

[Home]

Getting Started

R is a free software environment for statistical computing and graphics. It compiles and runs on a wide variety of UNIX platforms, Windows and MacOS. To download R, please choose your preferred CRAN mirror.

If you have questions about R like how to download and install the software, or what the license terms are, please read our answers to frequently asked questions before you send an email.

News

- R version 3.5.2 (Eggshell Igloo) prerelease versions will appear starting Monday 2018-12-10. Final release is scheduled for Thursday 2018-12-20.

- The R Foundation Conference Committee has released a call for proposals to host useR! 2020 in North America.

- You can now support the R Foundation with a renewable subscription as a supporting member

- R version 3.5.1 (Feather Spray) has been released on 2018-07-02.

- The R Foundation has been awarded the Personality/Organization of the year 2018 award by the professional association of German market and social researchers.

News via Twitter

News from the R Foundation

Figure A-1. The R Project for Statistical Computing

Installing Java

Download the Java installer (*https://java.com/download*) (see Figure A-2) and launch it for your Windows, Mac, or Linux platform.

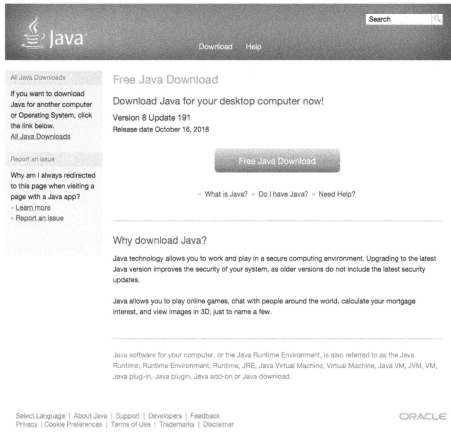

Figure A-2. Java download page

Starting with Spark 2.1, Java 8 is required; however, previous versions of Spark support Java 7. Regardless, we recommend installing Java Runtime Engine 8 (JRE 8).

If you're an advanced reader who is already using the Java Development Kit (JDK), note that JDK 9+ is currently unsupported. So you will need to downgrade to JDK 8 by uninstalling JDK 9+ or by setting JAVA_HOME appropiately.

Installing RStudio

While installing RStudio is not strictly required to work with Spark with R, it will make you much more productive and therefore, we recommend it. Download the RStudio installer (*https://www.rstudio.com/download*) (see Figure A-3), then launch it for your Windows, Mac, or Linux platform.

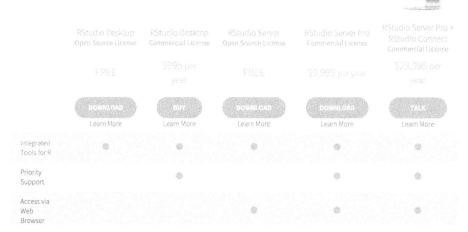

Figure A-3. RStudio download page

After launching RStudio, you can use its console panel to execute the code provided in this chapter.

Using RStudio

If you're unfamiliar with RStudio (see Figure A-4), you should make note of the following panes:

Console
 You can use a standalone R console to execute all the code presented in this book.

Packages
 This pane allows you to install `sparklyr` with ease, check its version, navigate to the help contents, and more.

Connections
 This pane allows you to connect to Spark, manage your active connection, and view the available datasets.

Figure A-4. RStudio overview

Chapter 3

Hive Functions

Name	Description
`size(Map<K.V>)`	Returns the number of elements in the map type.
`size(Array)`	Returns the number of elements in the array type.
`map_keys(Map<K.V>)`	Returns an unordered array containing the keys of the input map.
`map_values(Map<K.V>)`	Returns an unordered array containing the values of the input map.
`array_contains(Array, value)`	Returns TRUE if the array contains a value.
`sort_array(Array)`	Sorts the input array in ascending order according to the natural ordering of the array elements and returns it.
`binary(string or binary)`	Casts the parameter into a binary.
`cast(expr as a given type)`	Converts the results of the expression `expr` to the given type.
`from_unixtime(bigint uni xtime[, string format])`	Converts the number of seconds from Unix epoch (1970-01-01 00:00:00 UTC) to a string.
`unix_timestamp()`	Gets current Unix timestamp in seconds.
`unix_timestamp(string date)`	Converts a time string in the format yyyy-MM-dd HH:mm:ss to a Unix timestamp (in seconds).
`to_date(string timestamp)`	Returns the date part of a timestamp string.

Name	Description
`year(string date)`	Returns the year part of a date.
`quarter(date/timestamp/string)`	Returns the quarter of the year for a date.
`month(string date)`	Returns the month part of a date or a timestamp string.
`day(string date) dayof month(date)`	Returns the day part of a date or a timestamp string.
`hour(string date)`	Returns the hour of the timestamp.
`minute(string date)`	Returns the minute of the timestamp.
`second(string date)`	Returns the second of the timestamp.
`weekofyear(string date)`	Returns the week number of a timestamp string.
`extract(field FROM source)`	Retrieve fields such as days or hours from source. Source must be a date, timestamp, interval, or string that can be converted into either a date or timestamp.
`datediff(string enddate, string startdate)`	Returns the number of days from `startdate` to enddate.
`date_add(date/timestamp/ string startdate, tinyint/ smallint/int days)`	Adds a number of days to `startdate`.
`date_sub(date/timestamp/ string startdate, tinyint/ smallint/int days)`	Subtracts a number of days to `startdate`.
`from_utc_timestamp(\{any primi tive type} ts, string time zone)`	Converts a timestamp in UTC to a given time zone.
`to_utc_timestamp(\{any primi tive type} ts, string time zone)`	Converts a timestamp in a given time zone to UTC.
`current_date`	Returns the current date.
`current_timestamp`	Returns the current timestamp.
`add_months(string start_date, int num_months, out put_date_format)`	Returns the date that is `num_months` after `start_date`.
`last_day(string date)`	Returns the last day of the month to which the date belongs.
`next_day(string start_date, string day_of_week)`	Returns the first date that is later than `start_date` and named as `day_of_week`.
`trunc(string date, string for mat)`	Returns date truncated to the unit specified by the format.
`months_between(date1, date2)`	Returns number of months between dates `date1` and `date2`.
`date_format(date/timestamp/ string ts, string fmt)`	Converts a date/timestamp/string to a value of `string` in the format specified by the date format `fmt`.

Name	Description
if(boolean testCondition, T valueTrue, T valueFalseOr Null)	Returns valueTrue when testCondition is true; returns valueFal seOrNull otherwise.
isnull(a)	Returns true if a is NULL and false otherwise.
isnotnull(a)	Returns true if a is not NULL and false otherwise.
nvl(T value, T default_value)	Returns default value if value is NULL; else returns value.
COALESCE(T v1, T v2, …)	Returns the first v that is not NULL, or NULL if all vs are NULL.
CASE a WHEN b THEN c [WHEN d THEN e]* [ELSE f] END	When a = b, returns c; when a = d, returns e; else returns f.
nullif(a, b)	Returns NULL if a = b; otherwise returns a.
assert_true(boolean condition)	Throw an exception if condition is not true; otherwise return NULL.
ascii(string str)	Returns the numeric value of the first character of str.
base64(binary bin)	Converts the argument from binary to a base64 string.
character_length(string str)	Returns the number of UTF-8 characters contained in str.
chr(bigint double A)	
concat(string|binary A, string|binary B…)	Returns the string or bytes resulting from concatenating the strings or bytes passed in as parameters in order. For example, concat(foo, bar) results in foobar.
context_ngrams(array<array>, array, int K, int pf)	Returns the top-k contextual N-grams from a set of tokenized sentences.
concat_ws(string SEP, string A, string B…)	Like concat(), but with custom separator SEP.
decode(binary bin, string charset)	Decodes the first argument into a string using the provided character set (one of US-ASCII, ISO-8859-1, UTF-8, UTF-16BE, UTF-16LE, or UTF-16). If either argument is NULL, the result will also be NULL.
elt(N int,str1 string,str2 string,str3 string,…)	Return string at index number; elt(2,hello,world) returns world.
encode(string src, string charset)	Encodes the first argument into a BINARY using the provided character set (one of US-ASCII, ISO-8859-1, UTF-8, UTF-16BE, UTF-16LE, UTF-16).
field(val T,val1 T,val2 T,val3 T,…)	Returns the index of val in the val1,val2,val3,… list or 0 if not found.
find_in_set(string str, string strList)	Returns the first occurance of str in strList where strList is a comma-delimited string.
format_number(number x, int d)	Formats the number x to a format like '#,###,###.##', rounded to d decimal places, and returns the result as a string. If d is 0, the result has no decimal point or fractional part.
get_json_object(string json_string, string path)	Extracts JSON object from a JSON string based on JSON path specified, and returns JSON string of the extracted JSON object.
in_file(string str, string filename)	Returns true if the string str appears as an entire line in filename.

Name	Description
instr(string str, string substr)	Returns the position of the first occurrence of substr in str.
length(string A)	Returns the length of the string.
locate(string substr, string str[, int pos])	Returns the position of the first occurrence of substr in str after position pos.
lower(string A) lcase(string A)	Returns the string resulting from converting all characters of B to lowercase.
lpad(string str, int len, string pad)	Returns str, left-padded with pad to a length of len. If str is longer than len, the return value is shortened to len characters.
ltrim(string A)	Returns the string resulting from trimming spaces from the beginning (left side) of A.
ngrams(array<array>, int N, int K, int pf)	Returns the top-k N-grams from a set of tokenized sentences, such as those returned by the sentences() UDAF.
octet_length(string str)	Returns the number of octets required to hold the string str in UTF-8 encoding.
parse_url(string urlString, string partToExtract [, string keyToExtract])	Returns the specified part from the URL. Valid values for partToExtract include HOST, PATH, QUERY, REF, PROTOCOL, AUTHORITY, FILE, and USERINFO.
printf(String format, Obj… args)	Returns the input formatted according do printf-style format strings.
regexp_extract(string subject, string pattern, int index)	Returns the string extracted using the pattern.
regexp_replace(string INITIAL_STRING, string PATTERN, string REPLACEMENT)	Returns the string resulting from replacing all substrings in INITIAL_STRING that match the Java regular expression syntax defined in PATTERN with instances of REPLACEMENT.
repeat(string str, int n)	Repeats str n times.
replace(string A, string OLD, string NEW)	Returns the string A with all non-overlapping occurrences of OLD replaced with NEW.
reverse(string A)	Returns the reversed string.
rpad(string str, int len, string pad)	Returns str, right-padded with pad to a length of len.
rtrim(string A)	Returns the string resulting from trimming spaces from the end (right side) of A.
sentences(string str, string lang, string locale)	Tokenizes a string of natural language text into words and sentences, where each sentence is broken at the appropriate sentence boundary and returned as an array of words.
space(int n)	Returns a string of n spaces.
split(string str, string pat)	Splits str around pat (pat is a regular expression).
str_to_map(text[, delimiter1, delimiter2])	Splits text into key-value pairs using two delimiters. delimiter1 separates text into key-value pairs, and delimiter2 splits each key-value pair. Default delimiters are , for delimiter1 and : for delimiter2.

Name	Description
`substr(string binary A, int start)`	Returns the substring or slice of the byte array of A starting from start position till the end of string A.
`substring_index(string A, string delim, int count)`	Returns the substring from string A before `count` occurrences of the delimiter `delim`.
`translate(string|char|varchar input, string|char|varchar from, string|char|varchar to)`	Translates the input string by replacing the characters present in the `from` string with the corresponding characters in the `to` string.
`trim(string A)`	Returns the string resulting from trimming spaces from both ends of A.
`unbase64(string str)`	Converts the argument from a base64 string to BINARY.
`upper(string A) ucase(string A)`	Returns the string resulting from converting all characters of A to uppercase. For example, upper(*fOoBaR*) results in FOOBAR'\.
`initcap(string A)`	Returns string, with the first letter of each word in uppercase, and all other letters in lowercase. Words are delimited by whitespace.
`levenshtein(string A, string B)`	Returns the Levenshtein distance between two strings.
`soundex(string A)`	Returns soundex code of the string.
`mask(string str[, string upper[, string lower[, string number]]])`	Returns a masked version of str.
`mask_first_n(string str[, int n])`	Returns a masked version of `str` with the first n values masked. `mask_first_n("1234-5678-8765-4321", 4)` results in nnnn-5678-8765-4321.
`mask_last_n(string str[, int n])`	Returns a masked version of `str` with the last n values masked.
`mask_show_first_n(string str[, int n])`	Returns a masked version of `str`, showing the first n characters unmasked.
`mask_show_last_n(string str[, int n])`	Returns a masked version of `str`, showing the last n characters unmasked.
`mask_hash(string|char|varchar str)`	Returns a hashed value based on `str`.
`java_method(class, method[, arg1[, arg2..]])`	Synonym for reflect.
`reflect(class, method[, arg1[, arg2..]])`	Calls a Java method by matching the argument signature, using reflection.
`hash(a1[, a2…])`	Returns a hash value of the arguments.
`current_user()`	Returns current username from the configured authenticator manager.
`logged_in_user()`	Returns current username from the session state.
`current_database()`	Returns current database name.
`md5(string/binary)`	Calculates an MD5 128-bit checksum for the string or binary.
`sha1(string/binary) sha(string/binary)`	Calculates the SHA-1 digest for string or binary and returns the value as a hex string.

Name	Description
crc32(string/binary)	Computes a cyclic redundancy check value for string or binary argument and returns bigint value.
sha2(string/binary, int)	Calculates the SHA-2 family of hash functions (SHA-224, SHA-256, SHA-384, and SHA-512).
aes_encrypt(input string/ binary, key string/binary)	Encrypt input using AES.
aes_decrypt(input binary, key string/binary)	Decrypt input using AES.
version()	Returns the Hive version.
count(expr)	Returns the total number of retrieved rows.
sum(col), sum(DISTINCT col)	Returns the sum of the elements in the group or the sum of the distinct values of the column in the group.
avg(col), avg(DISTINCT col)	Returns the average of the elements in the group or the average of the distinct values of the column in the group.
min(col)	Returns the minimum value of the column in the group.
max(col)	Returns the maximum value of the column in the group.
variance(col), var_pop(col)	Returns the variance of a numeric column in the group.
var_samp(col)	Returns the unbiased sample variance of a numeric column in the group.
stddev_pop(col)	Returns the standard deviation of a numeric column in the group.
stddev_samp(col)	Returns the unbiased sample standard deviation of a numeric column in the group.
covar_pop(col1, col2)	Returns the population covariance of a pair of numeric columns in the group.
covar_samp(col1, col2)	Returns the sample covariance of a pair of a numeric columns in the group.
corr(col1, col2)	Returns the Pearson coefficient of correlation of a pair of a numeric columns in the group.
percentile(BIGINT col, p)	Returns the exact _p_th percentile of a column in the group (does not work with floating-point types). p must be between 0 and 1.
percentile(BIGINT col, array(p1 [, p2]…))	Returns the exact percentiles p1, p2, … of a column in the group. pi must be between 0 and 1.
percentile_approx(DOUBLE col, p [, B])	Returns an approximate _p_th percentile of a numeric column (including floating point types) in the group. The B parameter controls approximation accuracy at the cost of memory. Higher values yield better approximations, and the default is 10,000. When the number of distinct values in col is smaller than B, this gives an exact percentile value.
percentile_approx(DOUBLE col, array(p1 [, p2]…) [, B])	Same as previous, but accepts and returns an array of percentile values instead of a single one.
regr_avgx(independent, dependent)	Equivalent to avg(dependent).
regr_avgy(independent, dependent)	Equivalent to avg(independent).

Name	Description
regr_count(independent, dependent)	Returns the number of non-null pairs used to fit the linear regression line.
regr_intercept(independent, dependent)	Returns the y-intercept of the linear regression line—that is, the value of b in the equation dependent = a * independent + b.
regr_r2(independent, dependent)	Returns the coefficient of determination for the regression.
regr_slope(independent, dependent)	Returns the slope of the linear regression line—that is, the value of a in the equation dependent = a * independent + b.
regr_sxx(independent, dependent)	Equivalent to regr_count(independent, dependent) * var_pop(dependent).
regr_sxy(independent, dependent)	Equivalent to regr_count(independent, dependent) * covar_pop(independent, dependent).
regr_syy(independent, dependent)	Equivalent to regr_count(independent, dependent) * var_pop(independent).
histogram_numeric(col, b)	Computes a histogram of a numeric column in the group using b non-uniformly spaced bins. The output is an array of size b of double-valued (x,y) coordinates that represent the bin centers and heights.
collect_set(col)	Returns a set of objects with duplicate elements eliminated.
collect_list(col)	Returns a list of objects with duplicates.
ntile(INTEGER x)	Divides an ordered partition into x groups called *buckets* and assigns a bucket number to each row in the partition. This allows easy calculation of tertiles, quartiles, deciles, percentiles, and other common summary statistics.
explode(ARRAY a)	Explodes an array to multiple rows. Returns a row-set with a single column (col), one row for each element from the array.
explode(MAP<Tkey,Tvalue> m)	Explodes a map to multiple rows. Returns a row-set with a two columns (key,value), one row for each key-value pair from the input map.
posexplode(ARRAY a)	Explodes an array to multiple rows with additional positional column of int type (position of items in the original array, starting with 0). Returns a row-set with two columns (pos,val), one row for each element from the array.
inline(ARRAY<STRUCT<f1:T1, …,fn:Tn>> a)	Explodes an array of structs to multiple rows. Returns a row-set with N columns (N = number of top-level elements in the struct), one row per struct from the array.
stack(int r,T1 V1,…,Tn/r Vn)	Breaks up n values V1,…,Vn into r rows. Each row will have n/r columns. r must be constant.
json_tuple(string jsonStr,string k1,…,string kn)	Takes JSON string and a set of n keys, and returns a tuple of n values.
parse_url_tuple(string urlStr,string p1,…,string pn)	Takes URL string and a set of n URL parts, and returns a tuple of n values.

Chapter 4

MLlib Functions

The following table exhibits the ML algorithms supported in `sparklyr`:

Classification

Algorithm	Function
Decision trees	`ml_decision_tree_classifier()`
Gradient-boosted trees	`ml_gbt_classifier()`
Linear support vector machines	`ml_linear_svc()`
Logistic regression	`ml_logistic_regression()`
Multilayer perceptron	`ml_multilayer_perceptron_classifier()`
Naive-Bayes	`ml_naive_bayes()`
One vs rest	`ml_one_vs_rest()`
Random forests	`ml_random_forest_classifier()`

Regression

Algorithm	Function
Accelerated failure time survival regression	`ml_aft_survival_regression()`
Decision trees	`ml_decision_tree_regressor()`
Generalized linear regression	`ml_generalized_linear_regression()`
Gradient-boosted trees	`ml_gbt_regressor()`
Isotonic regression	`ml_isotonic_regression()`
Linear regression	`ml_linear_regression()`

Clustering

Algorithm	Function
Bisecting k-means clustering	`ml_bisecting_kmeans()`
Gaussian mixture clustering	`ml_gaussian_mixture()`
k-means clustering	`ml_kmeans()`
Latent Dirichlet allocation	`ml_lda()`

Recommendation

Algorithm	Function
Alternating least squares factorization	`ml_als()`

Frequent Pattern Mining

Algorithm	Function
FPGrowth	`ml_fpgrowth()`

Feature Transformers

Transformer	Function
Binarizer	`ft_binarizer()`
Bucketizer	`ft_bucketizer()`
Chi-squared feature selector	`ft_chisq_selector()`
Vocabulary from document collections	`ft_count_vectorizer()`
Discrete cosine transform	`ft_discrete_cosine_transform()`
Transformation using dplyr	`ft_dplyr_transformer()`
Hadamard product	`ft_elementwise_product()`
Feature hasher	`ft_feature_hasher()`
Term frequencies using hashing	`export(ft_hashing_tf)`
Inverse document frequency	`ft_idf()`
Imputation for missing values	`export(ft_imputer)`
Index to string	`ft_index_to_string()`
Feature Interaction transform	`ft_interaction()`
Rescale to [−1, 1] range	`ft_max_abs_scaler()`
Rescale to [min, max] range	`ft_min_max_scaler()`
Locality sensitive hashing	`ft_minhash_lsh()`
Converts to n-grams	`ft_ngram()`
Normalize using the given P-norm	`ft_normalizer()`
One-hot encoding	`ft_one_hot_encoder()`
Feature expansion in polynomial space	`ft_polynomial_expansion()`
Maps to binned categorical features	`ft_quantile_discretizer()`
SQL transformation	`ft_sql_transformer()`
Standardizes features using corrected STD	`ft_standard_scaler()`
Filters out stop words	`ft_stop_words_remover()`
Map to label indices	`ft_string_indexer()`

Transformer	Function
Splits by white spaces	`ft_tokenizer()`
Combine vectors to row vector	`ft_vector_assembler()`
Indexing categorical feature	`ft_vector_indexer()`
Subarray of the original feature	`ft_vector_slicer()`
Transform word into code	`ft_word2vec()`

Chapter 6

Google Trends for On-Premises (Mainframes), Cloud Computing, and Kubernetes

The data that created Figure 6-1 can be downloaded from *https://bit.ly/2YnHkNI.*

```
library(dplyr)

read.csv("data/clusters-trends.csv", skip = 2) %>%
  mutate(year = as.Date(paste(Month, "-01", sep = ""))) %>%
    mutate(`On-Premise` = `mainframe...Worldwide.`,
           Cloud = `cloud.computing...Worldwide.`,
           Kubernetes = `kubernetes...Worldwide.`) %>%
      tidyr::gather(`On-Premise`, Cloud, Kubernetes,
                    key = "trend", value = "popularity") %>%
      ggplot(aes(x=year, y=popularity, group=trend)) +
        geom_line(aes(linetype = trend, color = trend)) +
        scale_x_date(date_breaks = "2 year", date_labels = "%Y") +
        labs(title = "Cluster Computing Trends",
             subtitle = paste("Search popularity for on-premise (mainframe)",
                              "cloud computing and kubernetes ")) +
        scale_color_grey(start = 0.6, end = 0.2) +
        geom_hline(yintercept = 0, size = 1, colour = "#333333") +
        theme(axis.title.x = element_blank())
```

Chapter 12

Stream Generator

The `stream_generate_test()` function presented in Chapter 12 creates a local test stream. This function works independently from a Spark connection. The following example will create five files in the *source* subfolder. Each file will be created 1 second after the previous file's creation:

```
library(sparklyr)

stream_generate_test(iterations = 5, path = "source", interval = 1)
```

After the function completes, all of the files should show up in the *source* folder. Notice that the file sizes vary: this is so that it simulates what a true stream would do:

```
file.info(file.path("source", list.files("source")))[1]

##                         size
## source/stream_1.csv   44
## source/stream_2.csv  121
## source/stream_3.csv  540
## source/stream_4.csv 2370
## source/stream_5.csv 7236
```

The `stream_generate_test()` by default creates a single numeric variable Data-Frame.

```
readr::read_csv("source/stream_5.csv")

## # A tibble: 1,489 x 1
##        x
##    <dbl>
##  1   630
##  2   631
##  3   632
##  4   633
##  5   634
##  6   635
##  7   636
##  8   637
##  9   638
## 10   639
## # ... with 1,479 more rows
```

Installing Kafka

These instructions were compiled using information from the current Quickstart page of the official Kafka site (*https://kafka.apache.org/quickstart*). (Newer versions of Kafka will undoubtedly be available not long after this book is published.) The idea here is to "timestamp" the versions used in the example in "Kafka" on page 233:

1. Download Kafka.

   ```
   wget http://apache.claz.org/kafka/2.2.0/kafka_2.12-2.2.0.tgz
   ```

2. Expand the *tar* file and enter the new folder.

   ```
   tar -xzf kafka_2.12-2.2.0.tgz
   cd kafka_2.12-2.2.0
   ```

3. Start the Zookeeper service that comes with Kafka.

   ```
   bin/zookeeper-server-start.sh config/zookeeper.properties
   ```

4. Start the Kafka service.

```
bin/kafka-server-start.sh config/server.properties
```

Make sure to always start Zookeeper first, and then Kafka.

Index

H

H2O, 178-182
Hadoop Distributed File System (HDFS), 4
Hadoop YARN, 99, 122
HDInsight, 109, 128, 216
high-performance hardware, 93
histograms, 44
Hive project
 benefits of, 4
 functions, 255-262
 Spark integration with, 150
 SQL conventions, 39
homogeneous clusters, 216
Hortonworks, 101-103
httr package, 87
Hue, 112
hybrid cloud, 95, 112
hyperparameter tuning, 82

I

IBM Cloud, 108, 128
implicit partitions, 167
information storage and retrieval
 analog versus digital information, 2
 Apache Spark, 4-7
 cluster computing benefits, 8
 Google File System, 2
 Hadoop, MapReduce, and SQL, 3
 historical phases of, 2
 R computing language, 9-13
 sparklyr package, 13
input/ folder, 26
instances, 96

J

Java
 checking version installed, 17
 installing, 252
JDBC, 152
job execution, process of, 156
JSON file format, 146
Jupyter, 114

K

Kafka
 installing, 265
 reading and writing from, 227, 230
 streaming, 233

knitr package, 49
Kubernetes, 95, 111, 127, 264

L

lambda expressions, 200
large-scale data processing, 7, 202
Linux Cgroups, 100, 111
Livy, 114, 124, 216
local clusters, 18, 120
logistic regression, 4, 67, 82
logs and logging, 27, 130, 220

M

machine learning, 11, 54, 262-264
magrittr package, 14, 38
malformed entries, 140
map operation, 3, 200
MapR, 101, 103
MapReduce, 3
master parameter, 19
memory, 140, 171
Mesos, 100, 126, 160
Microsoft Azure, 109, 128
Microsoft HDInsight, 109, 128, 216
missing values, 37
mleap package, 88
MLlib, 262-264
modeling
 exploratory data analysis, 56-63
 feature engineering, 63
 overview of, 23, 46, 54
 pipeline models, 79
 predictive modeling, 54
 Spark streams, 230
 supervised learning, 66-72
 unsupervised learning, 72
multilayered perceptrons, 189

N

nodes, 96
NoSQL databases, 151

O

off-the-shelf hardware, 93
on-premises cluster computing
 cluster managers, 96-100
 overview of, 95
 Spark distributions, 101-103

About the Authors

Javier Luraschi is a software engineer with experience in technologies ranging from desktop, web, mobile, and backend, to augmented reality and deep learning applications. He previously worked for Microsoft Research and SAP and holds a double degree in mathematics and software engineering. He is the creator of `sparklyr`, `cloudml`, `r2d3`, `mlflow`, `tfdeploy`, and `kerasjs`.

Kevin Kuo builds open source libraries for machine learning and model deployment. He has held data science positions in various industries, including insurance, where he was a credentialed actuary. Kevin is the creator of `mlflow`, `mleap`, and `sparkxgb` among various R packages. He is also an amateur mixologist and sommelier.

Edgar Ruiz has a background in deploying enterprise reporting and business intelligence solutions. He is the author of multiple articles and blog posts sharing analytics insights and server infrastructure for data science. Edgar is the author and administrator of the `db.rstudio.com` website, and the current administrator of the `sparklyr` website. He also coauthored the `dbplyr` package and created the `dbplot`, `tidypre dict`, and `modeldb` packages.

Colophon

The animal on the cover of *Mastering Spark with R* is a rose fish (*Sebastes norvegicus*), a deep-sea species of rockfish. Although most rockfish live in the North Pacific, the rose fish is one of four species that live in the North Atlantic.

Juvenile rose fish live around fjords, while adults live in waters up to 3,280 feet deep. Adults average 18 inches long, although they can reach up to 3 feet. They are slow-growing fish that reach full maturity around 8 or 9 years old. Rose fish are viviparous; in winter, females give birth to 50,000–350,000 larvae. They can live to be 60 years old.

Rose fish are commercially fished, with an annual catch of 40 to 60 kilotons. They are considered to be severely overfished, especially given that it takes 8 or 9 years to reach reproductive age.

Many of the animals on O'Reilly covers are endangered; all of them are important to the world.

The cover illustration is by Jose Marzan, based on a black and white engraving from *Histoire Naturelle*. The cover fonts are Gilroy Semibold and Guardian Sans. The text font is Adobe Minion Pro; the heading font is Adobe Myriad Condensed; and the code font is Dalton Maag's Ubuntu Mono.

O'REILLY®

There's much more
where this came from.

Experience books, videos, live online
training courses, and more from O'Reilly
and our 200+ partners—all in one place.

Learn more at oreilly.com/online-learning

9 781492 046370